Preparing for Terrorism:
The Public Safety
Communicator's Guide

GEORGE BUCK, Ph.D.
LORI BUCK
BARRY MOGIL

THOMSON
DELMAR LEARNING

Australia   Canada   Mexico   Singapore   Spain   United Kingdom   United States

Preparing for Terrorism: The Public Safety Communicator's Guide
George Buck, Ph.D., Lori Buck, and Barry Mogil

Vice President, Technology and Trades SBU:
Alar Elken

Editorial Director:
Sandy Clark

Acquisitions Editor:
Mark Huth

Development:
Dawn Daugherty

Marketing Director:
Maura Theriault

Channel Manager:
Fair Huntoon

Marketing Coordinator:
Brian McGrath

Production Director:
Mary Ellen Black

Production Manager:
Larry Main

Production Coordinator:
Dawn Jacobson

Production Editor:
Elizabeth Hough

Editorial Assistant:
Jennifer Luck

Production Services:
TIPS Technical Publishing, Inc.

Cover Design:
Michael Egan

Library of Congress Cataloging-in-Publication Data

Buck, George.
  Preparing for terrorism: a public safety communicators guide / George Buck, Lori Buck, and Barry Mogil.
  p. cm.
Includes index.
  ISBN 1-4018-7131-3
  1. Terrorism—United States—Prevention. 2. Emergency management—United States. 3. Emergency communication systems—United States. I. Buck, Lori. II. Mogil, Barry. III. Title.
  HB6432.B819 2003
  363.3 ' 2--dc21

2003002347

ISBN: 1-4018-7131-3

CONTENTS

Introduction ix
About the Authors xi
Acknowledgements xv

Chapter 1 National and International Terrorism 1
Overview 1
Types of Terrorists 3
What Makes a Terrorist 3
Terrorist Targets 5
Recognizing Terrorist Events 5
An Overview of International Terrorism 7
Categories of Insurgent and Terrorism Groups 7
Categories of Risk 10
What This Book is About 11
Emerging Response Planning and Controversial Issues 11
Summary 13

Chapter 2 The Basic Concepts of Emergency Management 15
Overview 15
Emergency Management Agencies 15
Factors that Affect Emergency Management 17
Threats Facing the United States 19
Types of Threats 19
Potential Hazards 20
The Changing Context: A Window of Opportunity 21
The Emergency Manager 22
Communications Centers 24
Comprehensive Emergency Management 25
The Integrated Emergency Management System 27
Starting the Real World Planning Process 31
Hazard (Risk) Analysis 32

The Planning Process 33
Key Terms in Hazard Analysis Tools 33
Capability Assessment 34
Setting Goals 37
Getting Organized 38
Multi-hazard, All-hazard, Functional Planning 38
History of the ICS 38
Exercising 40
Summary.. 41

Chapter 3   Event Planning and Management 43
Overview 43
Crisis and Consequence Management 44
Incident Operations 45
Emergency Operation Plan (EOP) Components 51
Federal Response Team Missions and Functions 57
Summary 65

Chapter 4   Training 67
Overview 67
Introduction 67
What Training is Available? 69
How Do Communications Centers Get Training? 70
Reading and Self-study Courses 71
National Fire Academy 72
College and University Degrees 73
Government-sponsored Training Courses 74
Videotape Course 77
State and Local Domestic Preparedness Support Helpline 78
What Happens After Training? 80
Summary 81

Chapter 5   Facility Security 83
Overview 83
Facility Security Issues 83
Communications Centers are at Risk for Attacks 84
Emergency Planning and Drills 100
New Building Construction 100

Communications Tower Security 100
Questions 102
Summary 102

Chapter 6   Radio Interoperability 103
Overview 103
Interoperability: When Terrorism Strikes, Are You
    Prepared to Talk? 103
Technology 111
Questions 114
Summary 114

Chapter 7   Computer Security 115
Overview 115
Physical Security 116
Authorized Personnel 117
Local Access 119
Dial-up Access 119
Direct Connection to the Internet 120
Firewalls 121
Virtual Private Networks (VPNs) 121
Indirect Connection to the Internet 122
Assessing Security Risks 122
Redundant Systems 123
Redundant Data 124
Data Backup 124
Backup Center 124
Manual Preparation 124
Viruses 125
Increased Call Volume 125
Websites 126
Summary 126

Chapter 8   Personnel Needs 127
Overview 127
Committee Setup 131
Critical Incident Stress Debriefing (CISD) 131
Communications Recovery 136

Caregiver, How's Your Family? 137
Summary 139

Chapter 9   Joint Information Center/Media **141**
Overview 141
Joint Information Centers 141
Federal Government Role in JIC Operations 147
Headquarters-level Response Structure 150
Regional-level Response Structure 151
Continuing Actions 153
Summary 160

**Chapter 10   Wrapping it Up for Your
            Communications Center** **161**
The Role of the Communications Center Director 161
Additional Facility Safeguards 162
Common Sense in Handling Strangers 163
Get the Bigger Picture 164
A Note 166
Other Things to Bear in Mind When Planning 166
Training to Augment Your Facility's Security Plan 167
Other Points to Consider 168
Threats 173
Summary 179

Appendix A   Public Safety Precautions/Actions **181**
General Factors 181
Patient Management 184
Job Aid: First Response to a Terrorist Incident 189

Appendix B   Anthrax Threat Advisory **191**
Handling Anthrax and Other Biological Agent Threats 191

Appendix C   Indicators of Possible Agent Usage **195**
Indicators of Possible Chemical Warfare (CW) Agent Usage 195
Indicators of Possible Biological Weapon (BW) Usage 195
If Release is Suspected 196
Notification Essentials 196

Incident Objectives 196

Appendix D   Complete Agent Description 199
Anthrax 199
Brucellosis 200
Plague 202
Tularemia 203
Smallpox 204

**Appendix E   Internet Resources for
     Terrorism/ Disaster Planning** 207
Communications Center Links 207
Terrorism Links 207
Government Links 208
Academic and Institutions 209
CounterTerrorism Links 210
Biological Terrorism Links 211
Critical Infrastructure Protection Links 212
Anthrax Articles on the Web 213
Anthrax Websites 213
SmallPox Articles on the Web 213
SmallPox Websites 214
Plague Articles 214
Plague Websites 214
General Information 214
State Emergency Operation Plans 216

**Appendix F   Orange County Fire Rescue Emergency Communica-
     tions Center Standard Operating Procedure** 219
Section 6.7 PSAP Evacuation 219
Section 6.7 PSAP Evacuation Checklists 224

Appendix G   Overview of Community Actions 229
Communications Unit (9-1-1) 229
Communications Recommendations 231
Oklahoma City Bombing Final Report: Communications 232
City of Oklahoma City Communications 233

Appendix H   The Critical Infrastructure Protection Process
   Job Aid, Edition 1: May 2002 253
Table of Contents 253
I. Introduction 254
II. CIP Overview 255
III. CIP Process Methodology 257
IV. CIP Process Question Navigator 261
V. Infrastructure Protection Decision Matrix 262
VI. Establishing a CIP Program 263

Glossary 265
Index 273

INTRODUCTION

The making of this book started during the production of our very first book, *Preparing for Terrorism: An Emergency Service Guide*. It was at this time that discussion in the Buck household turned to the matter of communications centers being left out of response planning and training for terrorism events. Communications and logistics are the two areas in most events that seem to cause the most problems and have the most post-incident issues. After Oklahoma City and the first World Trade Center bombing, little was mentioned of the work of communications centers. Many people emerged as heroes during these tragedies, yet 9-1-1 communications staff who handled calls from the public and from radio traffic from the field were rarely mentioned. The stress of being a 9-1-1 telecommunicator is as real as any first responder. 9-1-1 operators are truly the first "first responders." We take calls from a frightened public and give them strength and comfort while emergency crews respond to their emergency. We do this behind the scenes, without much notice, and in the shadow of the more visible first responders—but this paradigm is starting to shift.

The terrorist attack on the Pentagon and the World Trade Center, as well as the hijacked airline crash in Pennsylvania, put a tremendous burden on 9-1-1 centers across the United States. The amount of calls in New York alone far exceeded any previous daily record. 9-1-1 operators handled calls from citizens trapped in the upper floors of the World Trade Center. Frantic callers told them not to hang up as the phone was their only lifeline. Some callers asked if they should jump. Imagine being on the line with a caller and being the last person to talk with them before the towers fell. Imagine a radio operator speaking with a firefighter or police officer as the towers fell. Most of us will never know the pain felt by 9-1-1 operators and communications staff in those areas on that day, but we all share a need to be heard by first responders. Our agencies need to stand up and take a role in emergency response planning and training for terrorist incidents. We need to be full partners in planning and coordination. We are communicators and we need to step up and communicate our needs, our abilities, our limitations, and our willingness to work towards a comprehensive emergency plan that provides the quickest and most effective response to our communities during a terrorist attack. As the first lines of George's first book, *Preparing for Terrorism: An Emergency Service Guide,* stated, it is not a matter of "if" but "when" the next World Trade Center event will occur. This book addresses information that communications center staff members need to know to help prepare for, respond to, and recover from a terrorist event. This event may not occur in their specific community, but as we saw during the anthrax attacks after September 11, 2001, citizens everywhere called 9-1-1 to report mysterious powder and suspicious packages. Many hazardous materials teams faced more calls during those three months than in the entire history of their units. This book covers information such as the history of terrorism, what role emergency management plays, how to help our personnel with personal needs during attacks, and information on computer systems and the protection of those systems. Interoperability is discussed by one of the leading specialists in the field. There is even a section of the book that gives a quick overview of the mission and function of various terrorism response teams that may respond to your emergency if requested. In the appendices of this book there is accurate information concerning weapons of mass destruction and indicators of their use. We have included (with permission from IFSTA) a section of the final report of the Murrah Federal Building bombing that covers the actions of communications staff. This section is valuable for showing exactly what occurred both during and after the event. This book is a start. It is not intended to answer all of the specific questions about specific systems in your communications center. It is designed to inform readers of what terrorism is, what it is capable of, and what resources will be available to you during an attack. It is designed to get communications directors thinking about how to support their staff during a large-scale crisis situation. This is a start—the rest is up to your agency and yourself to prepare for the worst and plan for a good response. One of the most compelling statements made shortly after the attack on the World Trade Center was from (then) Mayor Rudy Guiliani. He was asked what he had learned about his emergency service workers on September 11th that was new. Our hope is that every communications center's mayor or elected officials will someday have the same response—Mayor Guiliani's answer was "Nothing."

ABOUT THE AUTHORS

The following chapters were written or supplemented by contributing authors from various fields of expertise.

CHAPTER 4: TRAINING
Additional material was provided by Toni Edwards Finley

CHAPTER 6: RADIO INTEROPERABILITY
Written by Ron Haraseth

CHAPTER 7: COMPUTER SECURITY
Written by Gail Tyburski

CHAPTER 8: PERSONNEL NEEDS
Additional information was provided by Candy Grund

CHAPTER 9: THE JOINT INFORMATION CENTER
Additional material was provided by Bill Wade

CHAPTER 10: WRAPPING IT UP FOR YOUR COMMUNICATIONS CENTER
Written by Jennifer Hagstrom

DR. GEORGE BUCK

Dr. Buck has been involved in fire/rescue services and emergency management for more than 20 years. He is presently the deputy director/associate professor at the University of South Florida's Center for Disaster Management and Humanitarian Assistance. He teaches graduate-level courses specializing in emergency/disaster management and terrorism. Dr. Buck was previously at St. Petersburg College as Director of the Institute of Emergency Administration and Fire Science and at the National Terrorism Preparedness Institute. He has also served as a Fire Management Specialist with the United States Fire Administration in Emmitsburg, Maryland.

He previously served fourteen years in the Operations division of the Addison (Dallas County), Texas fire department. He was a principal member of the technical committee for "Emergency Management, NFPA 1600" and served on the committee from 1992 to 2000. He has spoken at many national and international conferences and has published articles, manuscripts, and white papers—both nationally and internationally. Dr. Buck is the author of Delmar's *Preparing for Terrorism: An*

Emergency Services Guide, and *Preparing for Biological Terrorism: An Emergency Services Guide*.

Dr. Buck's formal education includes an associate of science in fire science from the University of Hawaii, a bachelor of science in emergency administration and planning from the University of North Texas, and both masters and doctoral degrees in public administration.

LORI BUCK

Lori Buck has served in emergency services for twenty-two years. She was a firefighter for Kitsap County Fire District #1 and was a Public Fire and Life Safety Specialist with the Bremerton Fire Department. She is a graduate of and instructor for the National Fire Academy, located in Emmitsburg, Maryland. She has worked with organizations and schools all over the country as a private consultant for emergency planning for child-related facilities. She writes a monthly column for the Association of Public-Safety Communications Officials International, Inc. (APCO), an international 9-1-1 organization, and has published a children's workbook entitled *Impatient Pamela Says: Learn How to Call 9-1-1*. She has also published a book by APCO featuring past monthly public education articles.

Ms. Buck is currently the public education coordinator for Pinellas County 9-1-1 Emergency Communications, located in Clearwater, Florida. She is working to reduce the number of unnecessary calls to the 9-1-1 system. She works with all audiences, from young children to seniors, to educate them on the proper use of 9-1-1.

BARRY MOGIL

Barry Mogil worked for fifty-two years in emergency services. He started out as an ambulance driver and attendant for a local funeral home and rose through the ranks to become the administrative director of fire and EMS services for Pinellas County, Florida. He was then appointed director of emergency communications/9-1-1 for the county. Over the past fifty-two years, he has received a number of outstanding awards for public service. These awards include APCO Emergency Communications Director of the Year (1999), the Silver Hat Lifetime Achievement award, Florida NENA award (1998), Administrator of the Year—Florida Association of County EMS (1995), EMS Lifetime Achievement Award—State of Florida (1992), and many more. He has been an instructor for the Emergency Management Institute in Maryland, author of numerous articles, and a team leader for the Commission on Accreditation of Ambulance Services. He is currently a private consultant, and this will be his first major book.

RON HARASETH

DIRECTOR, AUTOMATED FREQUENCY COORDINATION, INC. (AFC)

AFC is the subsidiary of the Association of Public-Safety Communications Officials International, Inc. (APCO) that provides FCC licensing assistance through frequency coordination, RF engineering, and licensing preparation.

Mr. Haraseth has a thirty-year background in public safety land mobile radio. He is currently the director of AFC, which is certified by the FCC to process public safety frequency coordinations.

Prior to working for APCO, Mr. Haraseth worked for the State of Montana for twenty-seven years. He worked in the Department of Transportation Communications Bureau and the Department of Administration's Public Safety Communications office. Mr. Haraseth started as a field radio technician in 1972 and worked up through the ranks. He has held the positions of the local AASHTO frequency coordinator and local APCO frequency coordinator for the State of Montana. Mr. Haraseth was instrumental to the development and administration of Montana's highly regarded interoperability and mutual aid program.

Mr. Haraseth is a member of the Radio Club of America and an amateur radio operator (N7SKJ). Mr. Haraseth actively participated in the Public Safety Wireless Advisory Committee (PSWAC), and is currently involved with the FCC National Coordinating Committee (NCC).

GAIL TYBURSKI

Gail Tyburski began working with computers in 1982 at Pinellas Ambulance Service. There she was exposed to every facet of the ambulance service business and the ways that computers could enhance their operation. In January of 1988, she began working for EAI, Inc., a premiere software company providing software and support for police, fire, ambulance, and 9-1-1 agencies all over the United States and Canada. She advised and assisted the staff of more than 150 agencies in maintaining their systems and databases in all facets of the public safety business. Even though the data existing on these systems is public, the integrity of the data is of utmost importance. She is currently employed as a senior programmer/analyst for Pinellas County Emergency Communications, bringing her extensive national experience to Pinellas County's nationally recognized 9-1-1 service.

CANDY GRUND

Candy Grund has been a 9-1-1 operator for Pinellas County Emergency Communications for over nine years. She has been involved in 9-1-1 public education for the past five years. She is a graduate of National Fire Academy public education courses and has presented at national 9-1-1 conferences on topics relating to public education. She has co-written a children's 9-1-1 workbook published by Trellis

Publishing entitled *Impatient Pamela Says: Learn How to Call 9-1-1*. She provides public education for thousands each year and participates in SAFE Kids and the Pinellas County Critical Incident Stress Debriefing Team.

TONI EDWARDS FINLEY

Toni Edwards Finley has more than seven years of experience as a public safety telecommunicator and trainer in Gainesville, Florida. She left the dispatch floor three years ago to work for the Association of Public-Safety Communications Officials—International. She is the current editor of the association's monthly magazine, *Public Safety Communications/APCO Bulletin*, for which she has written extensively on training issues. You can reach her at *edwardst@apco911.org*. Ms. Finley is a graduate of the University of Florida and a United States Army veteran.

JENNIFER HAGSTROM

Jennifer Hagstrom started in public safety communications as a law enforcement dispatcher in 1985. She ultimately rose to the position of technical systems manager at the Alachua County Sheriff's Office in Gainesville, Florida, before leaving to take an editorial position in publications at APCO in 1998. At APCO, she frequently contributed articles and a regular column to its magazine, *Public Safety Communications/APCO Bulletin*. She currently holds the freelance position of contributing editor at APCO, and still works part-time in both administration and on the floor at her old dispatch center.

BILL WADE

Bill Wade has been involved with the fire, EMS, and life safety fields for over twenty-five years. He began as a volunteer in a fire department and first aid squad in the Philadelphia area in 1973. After four years as a medic in the Air Force, he settled in Tampa, Florida and worked for a private ambulance service as an EMT while going to paramedic school.

He worked briefly with the Clearwater, Florida fire department before being hired by Tampa in 1981. Besides working as a paramedic, Captain Wade has also been on staff at the Tampa Fire Academy and has been a member of the Hazardous Materials Response Team. Captain Wade has been a part-time faculty member in the EMS program at Hillsborough Community College since 1983.

Since 1995, Captain Wade has been serving as the public information officer for Tampa Fire Rescue. The Tampa Bay area is the fifteenth largest television market in the country. The area has four network television stations producing local news, a twenty-four hour local news station, a very active radio news market, and two daily local newspapers: The Tampa Tribune and the St. Petersburg Times.

ACKNOWLEDGEMENTS

It is very difficult to write a book as large as this and continue to work in our respective fields on a daily basis, travel, teach, and maintain our family life. We would like to first thank all of our children, who sometimes had to wait for us to finish chapters or meet with chapter authors. They have been patient and wonderful during all of the books we have written. So to Joshua, Nathan, Taylor, and George—thanks for understanding. To George's mom, Theresa Brennan, thanks for answering the phones, picking up Taylor from school, watching the kids when we had to travel, and always being there to help us. You have made our lives richer by being such a great part of our family. In teaching and training throughout the world, we have made some wonderful friends and we have learned as much from those friends as we hope we have shared with them. For all of you, we thank you for sharing yourselves with us.

There are many people who assisted with the formation and production of this book. The following is a list of people whom we thank for their contributions, assistance, and advice on this book:

Carrie Mahony (a great researcher), Lindsy Ingram, Jackie Weinreich, Darla Douglas, Donna Beim, Dave Byrom, Kathy Vacca, Bob Dipalma, C. J. Reid, Pam Montanari, Pat Modrak, Cindy Lorow, Mark Weinreich, Barry Luke, Bernie the NETC Resource Library researcher, Blaine County Commissioner Victor Miller, Alvin Roach (who kept our computers working throughout this project), Howard Douthit, Vicki Pegram, Steven Quigley, Toni Wyman, Patti Broderick, Candy Rostan, Richard Nowakowski, Don Howell, Ella Cora, Susan Walker, Gary Dempsey, Jim Lamont, Rick Webster, Terry Zimmerman, Thera Bradshaw, and (of course) the staff at Delmar—Dawn Daugherty and Betsy Hough.

—Dr. George Buck
—Lori Buck

There are many people who have helped me over the many years I have been employed in the public safety sector. There are a few, however, that even after retirement remain a constant influence on my ability to remain involved in public safety.

I would first of all like to thank my wonderful wife, soulmate, and best friend Shirley for the support and guidance she has given me for over fifty years. Our four children and our grandchildren also still support and encourage me. Lori Buck, who came into my life when I was but a short time away from retirement, has been a very positive influence and remains a counselor to me. I have learned a lot and very much appreciate having her both in my professional life and as a great friend. I also thank George Buck, whom I admire and appreciate as a scholar and friend. And finally, I'd like to thank the many friends and acquaintances who have had a positive influence on the successes I have enjoyed.

—Barry Mogil

CHAPTER 1

National and International Terrorism

OVERVIEW

This chapter will identify the basic response strategies to Weapons of Mass Destruction (WMD) events that build upon existing doctrine, while addressing the unique considerations of the terrorism environment and reviewing a brief history of terrorism. Communications staff members need to understand terrorism in order to help all first responders deal safely and effectively with a terrorist attack. This chapter will help you understand the risk of an attack on your center or in your community. You will be able to

- Define terrorism and the terrorist ideology
- Identify considerations for the selection of potential targets by terrorists
- Demonstrate an understanding of the difference between strategies and tactics
- Define terrorism

The United States Department of Justice defines terrorism as "the unlawful use of force against persons or property to intimidate or coerce a government, the civilian population, or segment thereof, in the furtherance of political or social objectives" (source: Federal Bureau of Investigation).

While this may be the official definition, the news media define terrorism in their own ways. A reporter in Southern California once made the statement that someone shooting out car windows on the freeway was committing "highway terrorism." With the collapse of ENRON, the news media has invented the label "economic terrorism." In researching this book, I have come across over 100 definitions for terrorism. This is why I have created a new definition of terrorism: *Terrorism is whatever the news media defines as terrorism for that day's news event.*

This suggests a broad definition of terrorism as an illegal act intended to cause a change in politics or social issues through the use of intimidation. To better understand terrorism, we must look at five commonly accepted variables.

THE VIOLENCE NEED ONLY BE THREATENED

Terrorist acts are designed to do one thing—instill fear. They make people feel vulnerable and worry that the government is unable, or unwilling, to provide adequate protection. A terrorist act need only be threatened to cause that level of fear, provided that the threat is perceived to be genuine and valid. Consider the impact on the United States airline and travel industries if a believable threat were made to bomb twenty United States planes over the next six months. If the public believed the threat, the resulting reduction in travel would be financially devastating to the industries and result in pressure on the government to respond to the terrorists' agenda. Recall, for example, the incident in the summer of 1986, when Middle Eastern terrorists threatened United States tourists in Europe with a campaign of terror and bombings. This simple threat reduced United States tourism almost forty percent and placed a severe economic burden on the European tourism industry.

FEAR IS THE ACTUAL AGENT OF CHANGE

For a terrorist to be successful, the population must be afraid. It is the population's fear that brings pressure for the change that the terrorist desires. As fear grows, distrust in the government's ability to protect the public increases. This distrust will result in either a policy change or an overthrow of the government. What if America's emergency first responders feared becoming targets? If responders became overly fearful and reluctant to act, public levels of panic would increase dramatically.

Sometimes, however, a terrorist plan will backfire. Sometimes a terrorist act is so reprehensible that it incites anger rather than fear. If terrorists cross the line from fear to anger, their agenda is unlikely to be furthered. Examples are fresh in our mind. The bombing of servicemen in a West German nightclub in the 1980s resulted in a strong military response against Libya for supporting the operations. More recently, the images burned into people's minds of the 168 people killed in Oklahoma City caused a reaction opposite to that desired by the perpetrator, Timothy McVeigh. The American public was not frightened but appalled by the act, and it rallied behind the victims and Oklahoma City as if in the common defense of a country at war. These strong, unyielding responses and sense of community do much to prevent terrorists from carrying out such acts.

TERRORISTS' VICTIMS ARE NOT NECESSARILY THE ULTIMATE TARGETS

Generally speaking, the actual victims (whether injuries or deaths) are not the specific targets of terrorist acts. Victims are only pawns in a terrorist's attempt to ignite fear in those who witness the attack. Many times, the victims are just in the right place (*right* meaning somewhere they feel comfortable and secure) at the wrong time.

THOSE WHO OBSERVE THE ACT ARE THE INTENDED AUDIENCE

Consider the media coverage that terror attacks generate. Extensive media coverage is a double-edged sword. On one hand, it helps pull the country or community together; on the other hand, it furthers the terrorist's agenda by allowing more people to witness, almost instantly, an event designed to instill fear.

A TERRORIST'S DESIRED OUTCOME IS POLITICAL OR SOCIAL CHANGE

A terrorist is trying to create enough individual fear or distrust of the government to force changes in social or political situations. Whether that change is to stop abortions, change a public policy, or get the public to relinquish a certain freedom or liberty (in order to give the government more power to protect them), the people who are fearful demand the change. A terrorist sets the terms for the cessation of hostilities. If we accept those terms, we must remember that we have allowed the terrorists to achieve their goal.

TYPES OF TERRORISTS

There are numerous ways to categorize or define terrorists—domestic or international, left or right wing, ideological, special interest, single issue, anarchists, neo-fascists, and so forth. We will discuss only a simple breakdown of the types of terrorists.

Domestic terrorists originate within the United States. More often than not, they hold extreme right-wing beliefs. This is not to say that domestic terrorists are never from the left wing politically (we will discuss left and right ideologies in just a moment), but right-wing terrorists comprise by far the largest and most active terrorist groups within the United States. Luckily, to this point, there is little organization among right-wing groups and they do not operate in concert. Recent meetings between some of the larger groups, however, are raising concerns that they are becoming better organized.

WHAT MAKES A TERRORIST

The continuum developed by Vetter and Perlstein in 1991 (see Figure 1-1) demonstrates how terrorist groups range from the radical far left to the reactionary far right. The major belief structure of the far left is that of the fair and equitable distribution of power, wealth, prestige, and privilege. This belief structure is described by many as the "Marxist" left, inasmuch as members of the far left believe in the writings of Karl Marx and therefore follow socialist or communist agendas.

These types of groups are more likely to engage in terrorist activities designed to prompt the public to allow the government greater power. This is accomplished by causing sufficient terror in the public that people demand that the government do more to protect them. This increased government involvement generally results in the reduction of individual liberties or freedoms in the interest of protection. Alternatively, the agenda could be the institution of more extensive social programs or the redistribution of wealth.

On the other end of the continuum is the reactionary far right, whose values are based on order and a binding and pervasive morality. The far right may include religious, separatist, or racial supremacy groups. Essentially these groups believe in less government intervention in social issues or, in many cases, no government intervention at all.

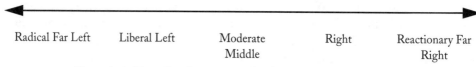

| Radical Far Left | Liberal Left | Moderate Middle | Right | Reactionary Far Right |

Figure 1–1 The political continuum developed by Vetter and Perlstein.

This is not to say that all left- or right-wing organizations are terrorist organizations. Other factors must be present before a group resorts to terrorist activities. In order for a group to be considered terrorist, it must meet the following three criteria.

AN EXTREMIST VIEWPOINT

Simply put, extremist viewpoints originate from belief in "one truth." That is to say, the group believes that there is only one answer to a particular debate, be it about abortion, sexual orientation, or another religious, social, or political issue. Holding an extremist viewpoint in itself by no means makes a group terroristic, of course. Many of us have one point of view on a specific topic—pro-life versus pro-choice is an excellent current example. Because the vast majority of people are tolerant of other points of view, however, they do not resort to terrorist activities, provided others do not force their points of view on them.

INTOLERANCE

It is when a group becomes intolerant of other points of view that it moves closer to the terrorist belief structure and meets the second criterion—intolerance. At this point, the group becomes unable to accept differences of opinion. This results in a belief that anyone who does not believe in the one truth is the enemy. Intolerance still does not make the group a terrorist group; one can be intolerant and still be a law-abiding citizen. It is the third and final criterion that distinguishes a terrorist.

A VILIFIED ENEMY

In the third criterion, people, governments, or countries with beliefs other than the group's are not only considered enemies, but also vilified. The enemy is seen as a hindrance to accomplishing the belief, as directly jeopardizing the one truth. Once this shift occurs, the enemy loses all value. The enemy is thus worthless and a direct threat to the individual's or group's belief structure. Any means necessary to defeat or overcome the enemy becomes acceptable. In other words, the end justifies the means. This can even be taken one step further to create the most dangerous of situations. If the terrorist accepts the belief that anyone who is not fighting the enemy *is* the enemy, then such people become worthless and may themselves be attacked regardless of age, gender, or relationship to the primary enemy.

TERRORIST TARGETS

Remember that terrorists want to instill fear in those who witness their attack. Therefore, they will look for a target that will give them as much media coverage as possible. The target may be a government facility if the terrorist is trying to portray the government as weak or inept (right-wing terrorist), or public facilities may be selected if the terrorist is trying to show the public that they need greater protection (left-wing terrorist). The terrorists of September 11th chose both.

Targets can be people, places, or infrastructure. People are targeted to make them fear that no place is safe. Places could be targeted because of historical or ideological significance or their value to the public. Infrastructure targets are attacked because they include those elements necessary for a community to function (e.g., roadways, bridges, water treatment plants). The more critical the infrastructure and the harder it is to restore its operations, the greater its potential for being targeted. For this reason, the terrorists targeted the World Trade Center (an economic symbol) and the Pentagon (a military symbol).

RECOGNIZING TERRORIST EVENTS

We mentioned earlier that terrorists may specifically target emergency responders. It is essential that responders understand how to recognize a terrorist event as early as possible so that changes can be implemented to maintain the tactical advantage. As soon as emergency responders recognize that an event is a terrorist attack, they must change the way they do business. They must avoid getting "blocked in," stay away from "choke points," use different response routes, implement security measures, be on the lookout for secondary and multiple devices, and so forth. To help identify terrorist events, responders need some tools for recognition.

There are two stages of terrorist event recognition. Pre-event recognition requires an awareness of conditions that might invite terrorism—watching out for people looking at the habits of emergency responders and communicating with local law enforcement officials about suspect groups. We will discuss this in the planning section of the text. Now we will focus on the other stage: the response phase.

THE RESPONSE PHASE

During the response phase, we need to be alert for indicators of terrorist incidents. These include

Location

- Occupancy
- Symbolism or history
- Site of public assembly
- Controversial nature of facility
- Infrastructure
- Critical nature of facility
- Vulnerability

Types of Events

- Bombings and Incendiary fires
- Events involving firearms
- Mass casualties without signs of trauma
- Epidemiological events

Conditions

- Ideal atmospheric conditions—unstable atmospheres with inversions and little air movement—which can be either natural (weather) or man-made (inside a building or a subway)
- Situations that appear to deliberately place people at a tactical disadvantage (choke points, high grounds, unexpected traffic diversions)

Timing of the Event

- Timed for maximum casualties
- Historic or significant dates

Other Observations

- Unusual casualty patterns or symptoms
- Odors
- Out-of-place containers or dissemination devices

AN OVERVIEW OF INTERNATIONAL TERRORISM

The trend in recent years has been a decrease in international terrorism. The number of international terrorist incidents fell from a high of 665 in 1987 to 304 incidents in 1997—up by eight incidents from the previous year but nevertheless reflecting a general downward trend. The intensity and types of explosives used (e.g., aircraft as guided missiles), however, has greatly increased.

Taken at face value, this is no indication of how serious the problem posed by terrorism has become. Deaths as a result of international and domestic terrorism appear to be increasing, although figures for so many parts of the world are elusive and inaccurate, largely because there is a problem of definition.

There is no universally accepted definition of terrorism, but for the purposes of this discussion we will use the following definitions:

Guerrilla Warfare

Military operations conducted by irregular troops against conventional government forces.

Insurgency/Insurgent

A rising or a revolt by a rebel or a revolutionary.

CATEGORIES OF INSURGENT AND TERRORISM GROUPS

During the 1970s and 1980s it was *de rigueur* to divide guerrilla and terrorist organizations by motivation, and such categories are still useful.

NATIONALIST GROUPS

Nationalist organizations were once regarded as the "aristocrats" of political violence and they remain a persistent and often deadly threat. Extremist Palestinian groups, such as the military wing of Hamas, the Provisional Irish Republican Army (PIRA), the Corsican separatist Front de la Liberation Nationale de la Corse (FLNC), and the Basque Euzkadi Ta Askatasuna (ETA), are clear examples of groups that have waged a sustained campaign of terror in an attempt to achieve a nationalist goal. They conduct isolated incidents of violence, often against civilian or economic targets. In Myanmar (formerly Burma), Sri Lanka, India, and Bangladesh, to name just a few, authorities have fought against insurgency campaigns, rural guerrilla warfare waged for a nationalist cause.

With the end of the Cold War and the removal of the shackles of East-West politics, new groups have emerged with their own nationalist agendas. In Chechnya, an insurgency campaign against the Russian Army humiliated Moscow and may ultimately pave the way for Chechnya's full independence. The newly independent republics of Georgia and Azerbaijan have also faced effective, damaging insurgencies by ethnic groups seeking to break free or join their kinsmen in another state.

In Europe, the breakup of Yugoslavia not only led to the continent's first "hot war" since 1945, but also spawned ethnic-nationalist terrorist groups in Macedonia and Kosovo.

In Africa, where tribal and ethnic divisions have frequently threatened the concept of statehood, fighting has broken out between rival ethnic groups in Rwanda and Burundi over their respective nationalist agendas.

"POLITICOS"

Left-wing terrorist groups—the Euro-terrorist axis of Action Direct, Brigate Rosse, Rote Armee Fraktion, and the Cellules Communistes Combattants were the most notorious examples—have largely declined in importance. Their aims were perhaps overambitious, and unlike nationalist groups, they did not have a ready pool of new recruits. Today, such organizations seem like an irrelevant part of the radical 1960s to the next generation, and with the end of the Cold War, Marxism has been discredited.

Nevertheless, left-wing ideology still has an important role to play among guerrilla groups such as Sendero Luminoso in Peru and Fuerzas Armadas Revolutionaries de Colombia (FARC) in Colombia. In 2002, the FARC used its influence of terror to force twenty-six mayors to resign in Colombia.

There has also been a decline in the number and influence of right-wing organizations, though they were never as numerous as those on the left.

RELIGIOUSLY MOTIVATED GROUPS

The most significant change in patterns of insurgency and political violence in the last twenty years has been the increase in the number of religiously motivated groups. Where secular ideologies have failed, a spectrum of radical religious organizations has sprung up. Near the end of the millennium, it was inevitable that cults such as the Branch Davidians and the Aum Shinrikyo would attract members—such organizations are often categorized as propagating "messianic terrorism." Even after the millennium, however, increasing numbers of political and nationalist movements are expressing a religious identity.

Ethnic or national divisions coinciding with indigenous religious beliefs is nothing new. For example, Muslim Kashmiris and Sikh Punjabis in Hindu India and Hindu Tamils in Buddhist Sri Lanka have rebelled. But religious radicalism—be it Christian, Islamic, Jewish, etc.—is increasingly setting the agenda in a number of conflicts and providing rebels with a new impetus. Religious extremist groups, sometimes referred to as subconflict organizations, aim to provoke full-scale insurrection.

For Islamic groups, the experience that outsiders gained fighting in the war against the Soviet Union in Afghanistan has proved invigorating; indeed, the emergence of "Afghanis" (foreigners, often Arabs or Pakistanis who fought alongside the various Mojahedin groups between 1980 and 1989) is one of the most important trends in international terrorism and insurgency violence. An estimated 10,000 Islamic mercenaries fought in Afghanistan. The irony is that they were indirectly equipped and trained by the CIA, via Pakistani Intelligence, and funded by Saudi Arabia. Conservative Arab governments encouraged young men to fight in Afghanistan to defend the honor of Islam.

In the valleys of Afghanistan, these young men were taught a firebrand message and given first-rate military training, including training in the use of sophisticated weapons and explosives. Once the Soviet Union retreated, the Afghanistan conflict became just another civil war, and thousands of these foreign recruits returned home to face unemployment. Today we are fighting some of the same people that the United States originally trained. In North Africa in particular, the return of those who had left to fight in Afghanistan proved disastrous for their native governments. In Kashmir, moderate insurgent groups like the Jammu and Kashmir Liberation Front (JKLF) have found themselves contending with Islamists determined to change their campaign from a struggle for self-determination into a jihad, or holy war.

Other Afghanis have used their experience in insurgency campaigns as far afield as the Middle East, India, China, Tajikistan, Bosnia, and the Philippines, or have sought renewed training in Sudan, Somalia, Lebanon, Iran, and Yemen. They are not mercenaries in the proper sense. They will not perpetrate acts of political violence on behalf of anyone, and their only cause is the Islamic revolution.

Christian extremist groups have made a particular impact in the United States. Christian fundamentalist, white supremacist, and extreme right-wing politics have converged, largely through the activity of anti-government militia groups and organizations such as the Ku Klux Klan. The United States has the dubious distinction of having suffered one of the worst domestic terrorist outrages in modern times—the Oklahoma City bombing—which was the work of at least one paranoid, anti-government former soldier with links to militia groups.

Such organizations and individuals are not in the habit of compromise. They are unlike politically mature organizations such as the Provisional Irish Republican Army (PIRA), which will usually steer clear of wholesale slaughter because it is invariably counterproductive. (PIRA learned this after the bombing of Christmas shoppers in Harrods, the Warrington bombing, and similar attacks cost the group sympathy in key constituencies, particularly in the United States.)

SINGLE-ISSUE GROUPS

Single-issue violence, sometimes referred to as "consumer terror," has also gained prominence with the growth of violent groups campaigning over issues such as animal rights, the environment, and abortion.

In the United States, federal agencies also talk of the "narcissistic terrorist," the loner whose deep sense of alienation pushes him to harbor a grudge and wage war on society. Theodore Kaczynski, the so-called Unabomber, was one such example.

SURROGATE TERRORISM

Finally, there are groups and individuals who are basically mercenaries—guns for hire. The most significant examples are the Abu Nidal Organization, the Japanese Red Army, which has abandoned its domestic motivations in favor of training other insurgency groups; and Carlos the Jackal, who is now serving time in a French prison for committing acts of terrorism. Organizations normally classified as politicos or nationalists may fit here as well if their motivation has become the pursuit of wealth.

CATEGORIES OF RISK

The following definitions categorize the levels of risk that terrorist, guerrilla, and insurgency groups pose to their governments in each of the states discussed in this document. The levels are not a measure of the number of deaths or casualties that might occur in each state or conflict. They indicate the threat to foreign interests and state security. The levels of risk are not an overall security rating for the country, but rather relate specifically to the groups in this document.

VERY HIGH RISK

- There is a state of civil war or widespread political or religious violence that undermines the effective running of the state and directly targets foreign interests or personnel.

- The government's authority does not extend across large parts of the country, and violence is the dominating factor in the operations of the state. The violence also directly threatens to bring down the government.

HIGH RISK

- There is an ongoing terrorist, guerrilla, or insurgency campaign in which foreign interests or personnel have been directly and indirectly targeted and where the authorities have no control in some significant area(s). It is possible to conduct normal daily life in less affected areas. The violence could indirectly bring down the government.

- The violence is a significant consideration in the government's planning and activities, but the state and administration are not under constant siege.

- The state is recovering from a full-scale guerrilla war or insurgency campaign.

MEDIUM RISK

- There is an ongoing terrorist, guerrilla, or insurgency campaign, but it is confined and does not threaten to undermine the state or its institutions.

- Foreign interests have been caught up in the violence, but they are not necessarily directly targeted and an attack against them is rare.

- The insurgency concerns a regional issue, but acts of violence may be conducted nationwide.

- The state is recovering from a high-risk insurgency campaign.

LOW RISK

- The government is fighting a terrorist, guerrilla, or insurgency campaign confined to remote geographical areas that can be easily avoided. The campaign does not pose a significant threat to foreign or domestic interests, although security forces will be involved in counterinsurgency activities.

- The insurgency concerns a regional issue and acts of violence are confined to that region.

MINIMAL RISK

- A state has experienced problems with terrorist groups but the threat has been effectively crushed. The state may have other security concerns, but it is not a major target of terrorist groups.

The United States government has a terrorism risk chart that should be placed in your communications center and updated as the Office of Homeland Security provides updates.

WHAT THIS BOOK IS ABOUT

This book focuses on the overall considerations and understandings of terrorism preparedness that communications center personnel must know. It is important, as a communications staff member, to understand that you are the lifeblood of emergency response. Understanding an Incident Command System (ICS), the local, state, and federal response plans, and your communications center protection actions will help both you and your first responders during a terrorist event. Your center is a critical piece of the response infrastructure and must be protected so that it will remain working for and communicating with all agencies who will arrive to assist. Very often communications personnel fail to fully participate in planning, training, and drilling simply because we have not been invited to do so, nor do we ask. Now is the time to do so, before something happens in your community (Figure 1-2).

EMERGING RESPONSE PLANNING AND CONTROVERSIAL ISSUES

The planning of emergency response procedures to terrorist events in the United States is a young and constantly changing set of ideas and information. As these events occur both domestically and internationally, we are constantly learning new and improved ways of managing the events and their consequences from other response organizations and military agencies. That reality means this book is dynamic in nature. The standard of care today may be outdated in the weeks or months to follow. It is important that communications centers constantly monitor changing information and standards, just as they monitor the changes in routine communications

Figure 1–2 Your center is a critical infrastructure—plan to protect it!

procedures. Refer to the Internet Annex for numerous sources considered authoritative by the National Fire Academy. Most of the resources are United States government and military sources. Other sources for up-to-date information include the Association of Public-Safety Communications Officials (APCO) and the National Emergency Number Association (NENA) or the Federal Emergency Management Agency (FEMA). All of these organizations can provide support and assistance to keep you current with the fast-changing world of communications.

SUMMARY

It is important that you understand the terrorist's motivations and identify potential targets of terrorist attacks in order to recognize terrorist events quickly. If you fail to have pre-event or event phase recognition in place, we are likely to find ourselves in over our head.

Know your community, the political and social conditions around you, likely targets, and how to recognize terrorist acts. Watch for patterns of 9-1-1 calls. Are many people calling with the same symptoms? Should a supervisor be notified? Should the health department be advised? Emergency communications is a valuable part of the emergency response system. The number of casualties we suffer is up to you, our leaders and the first responders.

QUESTIONS

1. List five commonly accepted variables of terrorism.

2. Define "reactionary far right."

3. What three criteria must a group meet to be considered a terrorist organization?

4. What are the two stages of terrorist event recognition?

5. What is the trend in international terrorism?

6. What is a nationalist organization?

7. Define a religiously motivated group.

8. List the five categories of terrorism risk.

9. What kind of targets do terrorist groups look for?

10. How many definitions can be found for terrorism?

CHAPTER 2

The Basic Concepts of Emergency Management

OVERVIEW

In this chapter, you will be introduced to the building blocks of an emergency management system. In any large-scale terrorist attack, an Emergency Operations Center (EOC) will be activated to handle all emergency functions. Communications center personnel must thoroughly understand what emergency management is and how it functions. Communication is key in any emergency, yet to function seamlessly understanding of both agencies must occur. The following topics are included:

- Comprehensive Emergency Management Systems
- Integrated Emergency Management Systems
- An overview of the Incident Command System
- An overview of emergency operational planning

This chapter covers basic emergency management information and will assist you in understanding the role emergency management plays in any large-scale terrorist attack.

EMERGENCY MANAGEMENT AGENCIES

Emergency management agencies over the years have taken on numerous roles and responsibilities for their communities. These have included emergency medical services, hazardous materials containment, and more recently, overall emergency management.[1] The fire services' role in emergency management is now a focal point in some agencies—it has become a necessity, not a luxury. The fire services' role is to review, focus on, and plan for an all-hazard emergency situation. This is not a new concept for fire services. They have often used many components of emergency management, such as command and control, incident command systems, and integrated emergency management systems in responding to emergencies and disasters.

Emergency management focuses on two basic tasks: preparing the organization to manage today's emergencies or disasters, and preparing for the new situations of tomorrow. The first responder leader who promotes knowledge of emergency management helps meet the department's responsibilities by improving the department's and the community's capability to prepare for disasters and to mitigate their effects. Taking a leadership role and making improvements in emergency management not only develops the department's ability to cope with tomorrow's disasters, but also improves the management of day-to-day operations. The fire department manages emergencies and disasters as part of an all-hazard system. This, in turn, will enhance the protection of its response personnel, increase its own professional reputation, and further protect the citizens it serves. Additional responsibilities may help to increase budgets and personnel. It will also be cost effective for many departments to inherit this activity instead of establishing new services. Naturally, a decision of this nature must be based on the jurisdiction's needs and its ability to provide and fund such services.

FIRE SERVICE INVOLVEMENT

In any emergency or disaster—small or large—involving fire department response, the goal is to mitigate the effects of the incident. A comprehensive emergency management system must be developed to assess potential situations and available resources, determine an appropriate proactive action plan, monitor the plan's effectiveness, and continually modify the plan to meet the realities of the situation.

If emergency service personnel are not functioning as part of an integrated emergency management system, their response (as well as that of fire, law enforcement, communications, and public works services) is effectively reduced—so too is the potential for communication and coordination with other agencies that may respond to the situation.

---

1. Emergency management is organized analysis, planning, decision-making, and assignment of available resources to mitigate (lessen the effect of or prevent), prepare for, respond to, and recover from, the effects of all hazards. The goal of emergency management is to save lives, prevent injuries, and protect property and the environment when an emergency or disaster occurs.

An effective emergency management system will benefit the community it serves in the following ways:

- Saving lives and reducing injuries
- Preventing or reducing property damage
- Reducing economic losses
- Minimizing social dislocation and stress
- Minimizing agricultural losses
- Maintaining critical facilities in functioning order
- Protecting infrastructure from damage
- Protecting mental health
- Lessening legal liability of government and public officials
- Providing positive political consequences for government action

Emergency management systems are useful at all locations, for all types of situations, and for all types of fire organizations. To be effective, emergency management systems must be suitable for use regardless of the type of jurisdiction or agency involvement. This may include single jurisdiction/single agency, single jurisdiction/multiagency, and multijurisdiction/multiagency involvement. The organizational structure must be adaptable to any disaster, applicable and acceptable to users throughout a community or region, readily adaptable to new technology, and capable of logical expansion from the initial response to the complexities of a major emergency.

Common elements in emergency organizational management, terminology, and procedures are necessary for maximum application of a system and use of existing qualifications and standards. In addition, it is necessary to quickly and effectively move resources committed to the disaster with the least disruption to existing systems.

FACTORS THAT AFFECT EMERGENCY MANAGEMENT

The most distinguishing issue of emergency management is the element of danger to lives and property. Over the past ten years, for example, an average of 130 firefighter fatalities and 100,000 injuries (accounting for the loss of at least one workday for each injured firefighter) have occurred each year (Figure 2-1). Every year, an average of 5,000 civilians die and 95,000 are injured in fire incidents. Property loss from 1989 to 1999 totaled $64 billion in direct costs, according to the Federal Emergency Management Agency (FEMA).

Untold numbers of people have been exposed to toxic materials, resulting in countless injuries and causing enduring pain and suffering. Among the health risks for firefighters and other rescue workers are hepatitis B and C, AIDS, and other infectious diseases. Cancer-producing chemicals are also a danger, for both firefighting forces and civilians.

Figure 2–1 Could anyone have imagined the damage and loss of life from September 11, 2001?

Emergency management is carried out in a constantly changing environment—although the situation may get better or worse, it seldom stays the same. The dynamics of such an environment present additional challenges to the Incident Commander (IC). The effectiveness of the incident action plan depends on a building's construction and contents—factors that may be difficult to determine. Danger increases due to flashover, backdraft, or the presence of hazardous contents. The dynamics of the incident may create difficulties in the gathering of accurate and current information, especially because of the limited time available at an incident scene. Additionally, emergency personnel reporting to the IC may not be able to adequately comprehend the overall situation.

A dynamic situation may require frequent shifts from offensive to defensive mode. For example, offensive modes include an aggressive interior attack and a direct attack on wildland fires. Defensive modes include an indirect attack on wildland fires, exposure protection, resource gathering, and the transition from offensive to defensive operations. Additionally, changes in priority may occur with regard to life, safety, incident stabilization, and property conservation.

In later reports of an incident, any compromise of firefighter and rescue worker safety, poor management of resources, or an inability to enlarge the command organization to meet the demands of the situation may have a negative impact on public perception of the emergency management department. Departments should be ready for any type of incident. Because there is no guarantee that adequate resources will be available for every incident, departments must be prepared to handle every incident—regardless of its size or complexity—with whatever resources are available.

THREATS FACING THE UNITED STATES

Every day the population of the United States is at risk from a broad spectrum of threats. These threats range from ordinary house fires to hazardous materials accidents on major interstate highways to natural disasters that affect many thousands of people. They also include the social threats posed by various forms of terrorism and civil disturbances.

The possible range of threats was brought home vividly during 1992, which ranked as one of the most devastating years in United States history (second only to the events of September 11th, 2001). Among the most notable incidents that year were the Chicago tunnel flood (a public works disaster), the series of earthquakes that occurred along California's San Andreas Fault, and the Los Angeles riots. The most significant threats in 1992, however, were posed by the hurricanes that battered states on both the Atlantic and Pacific coasts. Hurricane Andrew, which tore across Florida and Louisiana, resulted in fifteen deaths and billions of dollars in damages (an estimated $27 billion in Florida alone). Hurricane Andrew was probably the most costly natural disaster in United States history to date. It was followed by typhoons Omar and Brian in Guam and Hurricane Iniki in Hawaii that caused over a billion dollars in damages. The events of 1992 demonstrate the need to prepare for all potential hazards, both common and newly recognized. The cost of the attacks of September 11, 2001 is still being tallied.

TYPES OF THREATS

For the purposes of this book, there are two basic categories of threats: natural and man-made.

NATURAL THREATS

The largest single category of repetitive threats to communities comes from meteorological, geological, seismic, or oceanic events. They can pose a threat to any part of the country, and their impact can be localized or widespread, predictable or unpredictable. The damage resulting from natural disasters can range from minor to major, depending on whether they strike small or large population centers. Extremely severe natural disasters can have a long-term effect on the infrastructure of their location. Natural threats include avalanches, dam failures, droughts, earthquakes, floods, tropical storms or hurricanes, landslides, thunderstorms, tornadoes, tsunamis, volcanos, winter storms, and wildfires.

MAN-MADE THREATS

Man-made threats represent a category of events that has expanded dramatically throughout this century with advancements in technology. Like natural threats, they can affect localized or widespread areas, are frequently unpredictable, can cause substantial loss of life (besides the potential for damage to property), and can pose a significant threat to the infrastructure of their area. In this category are social threats that primarily come from actions by external, hostile forces against the property, population, or infrastructure of the government in the form of domestic civil disturbances. Other man-made threats include hazardous materials accidents at fixed facilities or in transport, power failures, structural fires, wildfires, telecommunications failures, transportation accidents, and terrorism.

RANKING OF THREATS

It is important to note that any ranking of a threat to communities and emergency services is potentially misleading because of:

- The wide variations that can occur with the application of different criteria to the same threat

- The significant differences that can occur in the impact of a particular threat on a region and the individual states within the region

- The fact that threats in one region are not necessarily applicable to another region

- Variances in the types of data collected on each threat

- The lack, in some cases, of available data with which to develop a reasonable ranking

POTENTIAL HAZARDS

The following potential hazards were identified by local emergency managers.[2]

- Hazardous materials incident—highway

- Power failure

- Winter storm

- Flood

- Tornado

- Drought

- Radiological incident—transportation

- Hazardous materials incident—fixed facility

- Urban fires

- Hazardous materials incident—rail

- Wildfire

2. *FEMA Capability Assessment and Hazard Identification Program for Local Governments,* 1992.

- Hazardous materials incident—pipeline

- Civil disorder

- Earthquake

- Air transport incident

- Dam failure

- Hazardous materials incident—river

- Rail transportation incident

- Hurricane/tropical storm

- Subsidence

- Radiological incident—fixed facility

- Nuclear attack

- Landslide

- Avalanche

- Volcano

- Tsunami

THE CHANGING CONTEXT: A WINDOW OF OPPORTUNITY

One of the most important structural changes for emergency management came with the end of the Cold War and the dissolution of the Soviet Union. While many uncertainties remain about the disposition of thermonuclear weapons formerly under the control of the U.S.S.R., public perceptions of the threat they pose have been significantly reduced—substantially altering the context of emergency management. The term "civil defense," for example, is properly applied to some programs that are relevant to all hazards. It has become so identified with preparedness for a nuclear attack, however, that a program with such a label is now much more difficult to justify in terms of size and resources.

While the perceived threat of nuclear war has diminished, low-profile threats posed by the proliferation of nuclear, chemical, and biological weapons are increasing. The chance of such deadly items falling into the hands of unstable or fanatical leaders has increased manyfold, but the world has been absent a galvanizing event such as that of September 11, 2001. The events of that day will have a long-term impact on the rewriting of disaster/terrorism preparedness programs. The national security emergency preparedness programs that have been the underpinning of emergency management are very difficult to justify when perceptions of threats have diminished, available revenues have declined, and demands for attention to domestic problems have increased instead.

One of the most dramatic changes in emergency management is the greater intrusiveness and influence of the news media. Disasters and emergencies provide dramatic news, and the appetite of the media—particularly television—is insatiable. This means that emergency management agencies often have to perform under intense media scrutiny. It also means that few emergencies and disasters will remain local—most will now be nationalized and politicized as the result of media coverage. This presents particular problems for maintaining the tradition that local and state governments take primary responsibility for emergencies while the federal government merely supplements their efforts. The media will pressure reluctant local and state leaders to "ask for federal help," will call on presidents to dispatch such help, and will urge representatives and senators to demand it on behalf of their constituents. This "CNN Syndrome" or "camcorder policy process" disrupts and distorts normal procedures and response patterns, making the best-laid plans and procedures vulnerable to disruption—indeed destruction—by one dramatic sound bite that the media can turn into a political shock wave.

The public expects more from all levels of government today—especially from the federal government. The reasons for this are not clear. They may stem from the general nationalization of the political system that has come with population mobility and the consolidation of the news media. It may also be that the general erosion of community, mutual aid, and self-help is resulting in people turning directly to the government for assistance in emergencies. An increase in expectations may fall upon the President, either because he is the only official elected by a national constituency and is the chief executive and commander-in-chief or because the President is the most visible symbol of our government. For people whose lives have been disrupted or who are in shock, symbols of competency and caring on the part of government are extremely important.

These changes in the context of emergency management present unprecedented challenges and opportunities. With memories of several disasters still fresh in people's minds, a change of administration, and renewed attention on the nation's domestic problems, government and private industry have the greatest opportunity in more than a decade to address and ameliorate the enduring problems of fire services and other emergency management agencies.

THE EMERGENCY MANAGER

A person given the job of emergency manager becomes leader and must look at all the components of an all-hazard comprehensive emergency system. That person has a job that is specifically defined by law. They are appointed and can be terminated by an elected mayor, a city manager, a county executive, or the board of commissioners. They must consider how the position's legal definition determines priorities. They also must consider their moral responsibility to save lives and reduce property damage, not only from the threat of fire but from all hazards facing their community. They may be well prepared to handle a fire emergency or disaster, but have no idea where to start planning for a comprehensive, integrated emergency management system. Where do the next higher levels of government come in? What can the emergency manager expect from county, state, and federal governments?

EMERGENCY MANAGER: ROLES AND RESPONSIBILITIES

An emergency manager has the responsibility of coordinating all components of the emergency management system in the jurisdiction. The components of an effective emergency management system include the fire service, law enforcement, elected officials, public works, parks department, emergency medical services, volunteers, other government agencies, the private sector, etc.

Emergency management involves knowing the possible threats to the community, planning for emergencies and disasters, operating effectively in a disaster, and conducting recovery operations after a disaster. Emergency management is the vital ingredient in the development of an effective emergency program. The emergency manager will become the key leader in planning, the coordinator of operations, the chief of staff to the city, the executive during an emergency response, the community liaison in building the emergency program, and the supporter of mitigation efforts.

Coordination of emergency services—public and private—is a matter of personal style. Frequent contacts with colleagues, sharing advice with other personnel, and combined training among agencies are all ways to make coordination easier. Most important of all, however, is knowing the boundaries of coordination. For example, coordination means that police and firefighters cooperate in setting up the security of a crowd-control line. The emergency manager should make certain that this responsibility is assigned and action is taken without conflict or controversy. The emergency manager, for example, should not tell a police chief how or where to set up security forces. The emergency manager assists in policy development for a time of disaster, not the development of standard operating procedures during a disaster.

In summary, the emergency manager serves the jurisdiction as the cement that holds together the various components of a mitigation, preparedness, response, and recovery program. The emergency manager's job is to draw together the other emergency response managers into an effective, coordinated response team. In addition, the emergency manager keeps a constant eye out for opportunities to avoid disasters through hazard mitigation efforts. In short, the emergency manager draws on a wide body of resources to produce the most effective emergency management system.

In the late 1970s, the Defense Civil Preparedness Agency—which later became a division of FEMA—supported a study of emergency management at state and local levels. The study was conducted by the National Governors Association, working closely with the federal reorganization project that created FEMA in 1979. The study called for a comprehensive emergency management policy and organizational structure at the federal level and articulated the concept of Comprehensive Emergency Management (CEM), including the description and fuller development of principles concerning the nature and relationship of the four phases of disaster management.

These events have greatly influenced the nation's emergency management environment. CEM is now widely accepted as a useful emergency management framework. Former FEMA Director James L. Witt, who was appointed by President Clinton, developed many proactive planning and management strategies, including CEM and the Integrated Emergency Management System (IEMS).

CEM and the IEMS are the cornerstones of all emergency management programs.

COMMUNICATIONS CENTERS

THE ROLE OF THE COMMUNICATIONS CENTER MANAGER

The primary responsibility of the communications center manager is to maintain a close liaison with emergency managers. In addition to the ongoing need to keep all public safety agencies in close, efficient communication, a constant stream of information from the EOC is vital. This keeps the manager aware of current conditions such as road closures, unsafe buildings, debris blockage, downed power lines, health hazards, biological or chemical threats, and any other unusual circumstances that might hinder or cause harm to responding units. Most of the ability to accomplish an efficient, well-operated communications system must be handled far in advance of an event. Protocols, general guidelines, and Standard Operating Procedures (SOPs) should be in place and practiced to make sure that the desired results can be accomplished.

SECURITY

Heightened security measures should be in place. Some suggestions include:

- All authorized personnel should display proper identification when entering and while on the premises.

- No visitors are permitted (including family members).

- Vendors or persons having business in the communications center must wear a visitor's ID badge or be escorted by a member of the staff while on premises.

- All mail and packages addressed to the center should be opened and monitored prior to entering the center.

- No unopened personal mail or packages may be brought onto the premises.

- During normal hours, business personnel must sign in through the administrative offices. No outside personnel are permitted after normal business hours without prior notice and authorization.

- Admittance of media personnel must be cleared by the designated Public Information Officer (PIO) or department head. Members of the media must show proper ID before being allowed into the communications center.

OTHER CONSIDERATIONS

Other considerations might include how to handle a cell phone call from outside the normal operating area, such as a call from a passenger on a high-jacked airliner. This might include asking for the flight number, departure and destination point, and approximate time of departure.

HANDLING CALLS

A predetermined plan to handle overloaded lines could be set up to automatically "spill" to secondary Public Safety Answering Points (PSAPs). A back-up, off-location site adds the ability to handle overloads or an alternative to handle calls if the primary site has to be evacuated. Additional notifications during a Weapons of Mass Destruction (WMD) event would include the FBI, state and county governments, and the health department. There should be a call triage system to categorize all incoming calls as emergency, essential, and non-emergency (to be handled at a later time or directed to another source) calls. A plan should establish who will handle calls normally dispatched by the PSAP (e.g., wheelchair transport, municipal buses, school buses, and limousines). A protocol should exist for how operators should handle response into a "hot" zone. A backup plan needs to be established to respond to the ongoing emergencies that will continue to occur outside of the major event. These include heart attacks, major trauma, burns, and other ongoing calls for assistance.

PERSONNEL

There should be a prearranged system for off-duty personnel to call in or come to a predetermined site. Provisions need to be made to house and feed personnel for the duration of the event and to care for the on-duty personnel. This system should include a plan for how operators are to handle response into a "hot" zone.

Another important duty is scheduling considerations. How long should operators be on shift without relief? Where will they stage when not at the console and how will they receive information about their families? These are but a few of the things a communications center manager must be prepared to deal with. That is why it is so necessary to preplan and to practice the plan.

COMPREHENSIVE EMERGENCY MANAGEMENT

CEM views emergency activities occurring in four separate but related phases: mitigation, preparedness, response, and recovery. These phases are visualized as having a circular relationship to each other. Each phase results from the previous one and establishes the requirements of the next one. The phases are related to the disaster by time and function, and each utilizes different personnel skills and management orientation. Activities in one phase may overlap those in the adjoining phases. For example, preparedness moves swiftly into response when disaster strikes. Response yields to recovery at different times, depending on the extent and kind of damage. Similarly, recovery should trigger mitigation—motivating attempts to prevent or reduce the potential of a future disaster.

MITIGATION

Mitigation refers to activities which reduce or eliminate the chance of a disaster's occurrence or the effects of a disaster. Recent research has shown that, while natural occurrences cannot be avoided, much can be done either to prevent major emergencies or disasters from happening or at least to reduce their damaging impact. Requiring protective construction to reinforce a roof, for example, will reduce damage from the high winds of a hurricane. Preventing construction in hazardous areas like flood plains can reduce the chance of flooded homes.

PREPAREDNESS

Preparedness is the planning for a response to an emergency or disaster and the work needed to increase the resources available to respond effectively. Preparedness activities help save lives and minimize damage by preparing people to respond appropriately when an emergency is imminent. To respond properly, a jurisdiction must have a plan, trained personnel, and the necessary resources. The objectives for local emergency management will describe the importance of a preparedness plan for every community and the value of human and material resources.

RESPONSE

Response activities are designed to provide emergency assistance to victims and to reduce the likelihood of secondary damage. The emergency services first responders and other support services are primary responders. Building and maintaining the capability to respond will be described in several sections of this book.

RECOVERY (PUBLIC)

Recovery is the final phase of the emergency management cycle. Recovery continues until all systems return to normal or near normal. Short-term recovery returns vital life support systems to minimum operating standards. Long-term recovery from a disaster may go on for years until the entire disaster area is completely redeveloped, either as it was in the past or for entirely new purposes that are less disaster prone. For example, portions of a town can be relocated and the area turned into open space or park land, thus providing the opportunity to mitigate future disasters. Recovery planning should include a review of ways to avoid future emergencies.

RECOVERY (PRIVATE)

One area that is often overlooked by communities is the recovery of the private sector. If local government does not assist in the recovery of businesses, the community loses tax revenue. Later chapters will lay out a working knowledge of a business continuation plan and will assist local government officials in the development of a public/private partnership.

The common factors among all types of natural and technological disasters, including terrorism, indicate that many of the same disaster management strategies can apply to all kinds of hazards. These common management approaches are a principal component of CEM.

The burden of disaster management, and obtaining the resources for it, require a close working partnership among all levels of government (local, state, and federal) and the private sector (business and the public). This final part of CEM calls for a conscientious effort to draw on the widest possible range of emergency management resources.

THE INTEGRATED EMERGENCY MANAGEMENT SYSTEM

IEMS is a management tool that reduces property damage, prevents injuries, and saves lives. The system has been implemented and tested time after time with great success. FEMA developed the system in 1983 to improve the ability of emergency management personnel to respond to major emergencies and disasters. Past successful implementations of the system are documented throughout the United States. IEMS has been used on a barge collision and fire in Tampa Bay, Florida, a train derailment in Louisiana, and (most recently) the Oklahoma City bombing.

Agencies using IEMS as an effective emergency management tool realize that no single agency can adequately respond to a major disaster. When public and private groups agree that emergency preparedness and efficiency are of vital importance, development of IEMS has begun. IEMS is the first step toward working together to protect emergency responders and the community they serve.

THE IEMS OVERVIEW

Emergency management[3] and planning are essential for effective disaster response. IEMS emphasizes teamwork in preparing for and responding to disasters, and offers a conceptual framework for organizing and managing emergency protection efforts.

An integrated approach to emergency management incorporates all available resources for responding to man-made and natural emergencies or disasters, as well as the full range of issues relating to the four phases of a CEM system (see previous section). Once the system is in place, it provides a means of efficiently incorporating resources from the private sector and other levels of government.

Every local area has distinct groups with differing emergency response capabilities. These groups include program management, super and tactical EOCs, law enforce-

---

3. The term "emergency management" as used in the United States equals the terms "disaster management" and "civil defense" as defined by the United Nations Department of Humanitarian Affairs.

ment, emergency medical services and fire departments, voluntary organizations such as the American Red Cross or Salvation Army, and many others. When these resources are linked through planning, direction, and coordination, and have clearly defined roles and functions, they are components of an IEMS.

Carry that integrated approach into the larger universe of regional, state, and federal resources and support relationships, and an IEMS is established. This can be achieved through mutual support with other jurisdictions, good lines of communication with other governmental levels, and dual use of emergency management resources.

Outside factors that affect the system include the hazard or emergency to be faced and associated political, social, and economic issues.

THE IEMS CONCEPT

The goal of the IEMS is to develop and maintain a credible emergency management capability throughout the United States by integrating activities along functional lines at all levels of government and—to the fullest extent possible—across all hazards.

Agencies can begin to achieve this goal by doing the following:

- Establishing an emergency management program tailored to the specific requirements of each community

- Developing plans that consider functions common to all types of emergencies, as well as which responses are unique to specific types of emergencies

- Incorporating existing personnel assignments, operating procedures, and facilities used for day-to-day operations into the disaster response plan

- Integrating emergency management planning into national and local government policy-making and operations

- Encouraging the operational use of an Incident Command System (ICS) to incorporate government functions (including law enforcement, public works, environmental protection, fire protection) and elected officials into the emergency management system

- Incorporating all government jurisdictions (i.e., surrounding counties and other cities) into the emergency management system

- Establishing close working relationships with private-sector organizations, such as the American Red Cross, construction contractors, associations, and private search and rescue teams, etc.

- Addressing all phases of a CEM system: mitigation, preparedness, response, and recovery (short- and long-term, public and private)

THE IEMS PROCESS

IEMS has been introduced into the United States network of emergency management organizations representing thousands of jurisdictions that face different kinds of hazards and require different capabilities and skills. Implementing the IEMS process requires different levels of effort by each jurisdiction and results in the identification of different functional areas requiring attention. The process is logical and applicable to all jurisdictions regardless of size, level of sophistication, potential hazards, or current capabilities.

In order to provide a complete description of the IEMS process, each step is described below as it would apply to a jurisdiction that has done little toward developing the required capability, given its potential hazards.

Although IEMS emphasizes capability development, the process recognizes that current operations must be conducted according to existing plans with existing resources, and that these operations can contribute to the developmental effort. Therefore, the process includes two paths: one focusing on current capabilities and activities (Steps 1-7), and the other emphasizing capability improvement (Steps 8-13).

1. *Hazards Analysis.* Essential ingredients for emergency planning include knowing what could happen and the likelihood of it happening, and being aware of the magnitude of the problems that could arise. The first step, then, is for the jurisdiction to identify the potential hazards and to determine the impact each of those hazards could have on people, property, and the environment. This task need not be complicated or highly sophisticated to provide useful results. What is important is that any hazard that poses a potential threat to the jurisdiction be identified and addressed in the jurisdiction's emergency response planning and mitigation efforts.

2. *Capability Assessment.* The jurisdiction must assess its current capability for dealing with the hazards that have been identified in Step 1. Current capability is determined by the standards and criteria FEMA has established as necessary to perform basic emergency management functions (e.g., alerting and warning, evacuation, and emergency communications). This not only provides a summary of the capabilities that exist, upon which current plans should be prepared (Step 3), but also leads to the identification of the jurisdiction's weaknesses (Step 8).

3. *Emergency Operations Plan.* A plan should now be developed with functional annexes common to the hazards identified in Step 1. Those activities unique to specific hazards should be described separately. This approach is a departure from previous instruction, which stressed development of hazard-specific plans. Existing plans should be reviewed and modified to ensure applicability to all hazards that pose a potential threat to the jurisdiction. The exact format of the plan is less important than the assurance that the planning process considers each function from a multi-hazard perspective.

4. *Capability Maintenance.* Once developed, the ability to take appropriate and effective action against any hazard must be continually monitored or it will diminish significantly over time. Plans must be updated, equipment must be ser-

viced and tested, personnel must be trained, and procedures and systems must be exercised. This is particularly important for jurisdictions that do not experience frequent large-scale emergencies.

5. *Mitigation Efforts.* Mitigating the potential effects of hazards should be given high priority. Resources utilized to limit a hazard or its effects can minimize loss and suffering in the future. For example, proper land-use management and stringent building and safety codes can lessen the effects of future disasters. Significant mitigation efforts can also reduce the level of capability needed to conduct recovery operations, thereby reducing the capability shortfall that may exist. The results of these efforts will be reflected in future hazards analyses (Step 1) and capability assessments (Step 2).

6. *Emergency Operations.* The need to conduct emergency operations may arise at any time. The operations must be carried out under current plans and with current resources, despite the existence of plans for making improvements in the future. Such operations can provide an opportunity to test existing capabilities under real conditions.

7. *Evaluation.* The outcome of the emergency operations (Step 6) should be analyzed and assessed in terms of actual *versus* required capabilities and considered in subsequent updates of Steps 2 and 8. Identifying the need for future mitigation efforts should be an important part of each evaluation. Tests and exercises should be undertaken for the purpose of evaluation, especially in areas where disasters occur infrequently.

8. *Capability Shortfall.* A capability shortfall is the difference between current capability (Step 2) and the optimum capability reflected in the standards and criteria established by policy. The areas not currently meeting the assessment criteria should receive primary consideration in the jurisdiction's multiyear development plan (see Step 9).

9. *Multi-Year Development Plan.* Based on the capability shortfall identified in Step 8, the jurisdiction should prepare a multiyear development plan tailored to meet its unique situation and requirements. The plan should outline what needs to be done to reach the desired level of capability. Ideally, this plan should cover a five-year period so that long-term development projects can be properly scheduled and adequately funded. The plan should include all emergency management projects and activities to be undertaken by the jurisdiction regardless of the funding source.

10. *Annual Development Increment.* Using the multiyear development plan as a framework for improving capability over time, the actions to take for the upcoming year should be determined in detail. Each year, situations change and what was accomplished the previous year may have exceeded or fallen short of expectations. These factors should be reflected in modifications to the multi-year development plan and the determination of next year's annual increment.

Through this process, emergency managers can provide local and national officials with detailed plans for the coming year and requirements for financial and technical assistance in support of these efforts.

11. *State and Local Resources.*[4] State and local governments are expected to continue their own capability development and maintenance efforts as they have done in the past. Some activities identified in the annual increment may be accomplished solely with local resources, while others may require state and/or federal support. Whatever the source of funding and other support, each project and activity should represent a necessary building block in the jurisdiction's overall capability development program.

12. *Federal Resources.* The federal government continues to provide policy and procedural guidance, financial and technical support, and staff resources to assist state and local governments in developing and maintaining capability. FEMA's Comprehensive Cooperative Agreement with states will remain the vehicle for annually funding for FEMA-approved projects throughout the United States and its territories.

13. *Annual Work Increment.* As capability development projects and activities are completed and the results of the process are reviewed, the jurisdiction's capability assessment shortfall (Steps 2 and 8) will change. Emergency operation plans should be revised to incorporate these improvements. Multi-year development plans also should be modified in view of these changes and the experience gained through exercises and actual emergency operations. Each state should provide a method for recording and consolidating local annual work increments.

STARTING THE REAL WORLD PLANNING PROCESS

The ability of a community to respond and mitigate a large-scale emergency depends on its planning process. The CEM and IEMS steps overlap and are repeated many times. Emergency managers are constantly reexamining their needs, reassessing their goals, planning actions, and evaluating results. The disaster plan, therefore, is a living, functional document and must be maintained throughout the years.

The first step in the planning process, as stated in the previous section, is to identify the problems and needs faced by the community. Every community is different, both in terms of the danger it faces and the resources it has available. For an emergency manager, one of the first tasks is to determine the potential for emergency situations to occur in the community and to evaluate the ability to respond. This is accomplished by utilizing the first two steps in the IEMS process: hazard analysis and capability assessment.

---

4. The three levels of government in the United States are local (city, township, village, county, etc.), state (State of New York, State of Washington, State of Hawaii, etc.), and federal (Washington, D.C.).

HAZARD (RISK) ANALYSIS

A good way to determine what may happen in the future is to review the past. Therefore, the first step in conducting a hazard analysis is to review a community's history of disasters and major emergencies and determine which were caused by hazards still present in the community. Planners should talk to citizens who know local history and teachers in local high schools, community colleges, and universities to find out whether any major emergencies or disasters occurred before the records of your planning department were initiated.

The collection and analysis of this information helps the emergency manager ascertain how vulnerable the community is to different kinds of disasters, and establishes part of the basis for making resource allocation decisions. The hazard analysis system is essentially a process for using a common set of criteria to determine and compare the risks that the community faces. Hazards are rated and scored in a way that allows for easy comparison.

The rating and scoring system involves the following four criteria, which can each be rated as *low, medium,* or *high* for each identified hazard.

- History—the record of previous disasters in the community
- Vulnerability—the number of persons who might be killed or injured, and the value of property that might be destroyed or damaged
- Maximum threat—the worst-case scenario of a hazard; that is, the set of circumstances in which the emergency will have greatest impact
- Probability—the likelihood that the disaster will occur

To perform a hazard analysis we must first review the elements of an emergency and then consider the cascade effect (discussed in the next section). Elements of an emergency include

- Probability
- Vulnerability (effect on the community)
- Predictability (likelihood of the emergency's occurrence)
- Frequency
- Controllability
- Speed of onset
- Duration
- Protection action options
- Scope/intensity
- Sources of assistance

This analysis is necessary to identify the resources needed to manage these hazards. It also allows us to plot our hazards on a community map, set planning priorities, and develop response models.

THE PLANNING PROCESS

The planning process includes the following steps:

- Identify problems and needs
- Set goals
- Determine objectives
- Set priorities
- Design action programs
- Evaluate results

The objective is to systematically identify and analyze the natural and technological hazards that threaten the jurisdiction, including the new threat of domestic terrorism, and use the results as a basis for multiyear program planning (see the thirteen steps detailed in the section on the IEMS process). Emergency planning should be based on those hazards that pose potential threats and significant consequences to the local jurisdiction. It is vital to understand the nature and implications of the hazards to which the population is, or may become, vulnerable.

The second part of a hazard analysis is to develop a cascade effect model. The cascade effect occurs when one emergency triggers others. For instance, what other kinds of hazards might be triggered by a truck bomb? By recognizing the cascade effect and relating this phenomenon to community hazards, the emergency manager can identify several primary hazards capable of triggering a wide range of other hazards (see Figure 2-2).

KEY TERMS IN HAZARD ANALYSIS TOOLS

ELEMENTS AT RISK

- Population
- Structures
- Public services
- Continuity of government
- Economy
- Political activities

EMERGENCY / DISASTER

- Hazard/hazard agent

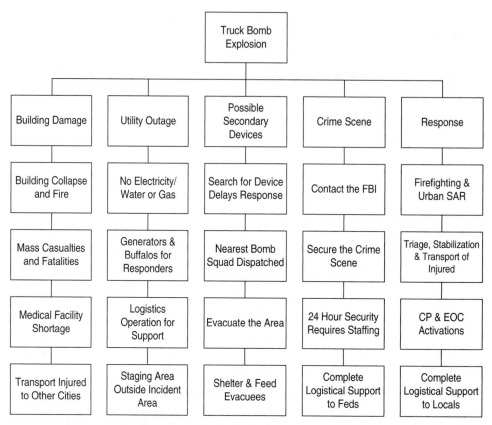

Figure 2–2 The Cascade chart.

- Incidence

- Local conditions

- Physical properties

- Probability

- Risk

- Risk reduction measures

- Vulnerability

CAPABILITY ASSESSMENT

Hazards analysis describes the dangers that a community may face. A capability assessment describes the likelihood that a community will deal well with those dangers. The following list includes the functional areas and the major elements within each area for which assessment criteria have been developed. A review of this list will provide a snapshot of the activities and resources that an effective emergency management organization should be able to provide.

EMERGENCY MANAGEMENT ORGANIZATION

- Legal authority
- Budget development
- Selection and training of the coordinator and staff
- Hazards analysis and multi-year development
- Written agreements for aid and resources
- Private-sector support

EMERGENCY OPERATIONS PLANNING

- Responsibilities
- Plan components
- Plan content
- Approval and promulgation
- Plan distribution
- Plan maintenance

RESOURCE MANAGEMENT

- Evaluation of needs
- Planning and preparedness
- Timely and effective utilization

DIRECTION AND CONTROL

- State and local facilities
- Assignment of responsibilities
- Protection
- Emergency operations
- Life support
- Damage assessment
- Maintenance

EMERGENCY COMMUNICATIONS

- Primary emergency communications
- Backup emergency communications

STAFFING AND TRAINING

- Protection

- Emergency power
- SOPs
- Emergency Broadcast System
- Maintenance

ALERTING AND WARNING SYSTEMS

- Point of contact for official information
- Warning points
- Staffing and training
- Emergency power
- SOPs
- Special locations and arrangements
- Maintenance

EMERGENCY PUBLIC INFORMATION

- Point of contact for official information
- Coordination and authoritative spokesperson
- Development and distribution of Emergency Public Information (EPI) materials
- Rumor control
- Media
- Reentry
- Emergency Broadcast System

CONTINUITY OF GOVERNMENT

- Lines of succession
- Safeguarding essential records
- Predelegation of authority
- Alternate headquarters
- Protection of government resources

SHELTER PROTECTION

- Congregate lodging facility
- Standard public shelters
- Shelter upgrading
- Expedient shelters
- Stocking

- Marking

- Reception and care

- Shelter managers

EVACUATION

- Preparation

- Movement

PROTECTIVE MEASURES

- Prevention of individual exposure to hazards

- Mitigation of impending or actual exposure

- Classification system and action level system

- Supply and maintenance of protective devices

- Procedures for recovery and reentry

EMERGENCY SUPPORT SERVICES

- Preparedness

- Law enforcement

- Fire and rescue

- Public health

- Transportation

- Medical services

- Public works

- Utilities

EMERGENCY REPORTING

- Monitoring

- Collection

- Processing

- Analysis

- Dissemination

SETTING GOALS

Determining objectives is the first step in developing an Emergency Operations Plan (EOP). In this plan the emergency manager begins to decide what he wants to accomplish in terms of the type and extent of emergency management capabilities, and develops a timeframe for accomplishing these goals.

The objective of all emergency management functions is to develop and maintain a comprehensive EOP based on hazard analysis, existing resources, and current operational capabilities in order to deal effectively with any kind of emergency or disaster.

GETTING ORGANIZED

To a great extent, the abilities of emergency management, a communications center, and the community to respond quickly and effectively to a large-scale emergency depend upon the quality and scope of the planning process.

In some ways the phrase "planning process" is misleading because it may imply that planning is a one time effort performed in advance of a disaster operation. People sometimes think that once a document titled "Disaster Plan" has been produced, the process is over. On the contrary, the plan itself may be less important than the process that produced it.

Plans and SOPs grow old and can quickly become outdated. The planning process, on the other hand, is ongoing. Planning is as much an attitude regarding management orientation as it is procedures that make up a document. Thus, in a real sense, planning is never finished.

An EOP is a document that contains information about actions designated by the emergency manager that may be taken by a governmental jurisdiction to protect people and property in the time of a natural or man-made disaster or the threat of such a disaster. An EOP also details the tasks that are to be carried out by specified organizational elements (public and private) during a disaster, based on established objectives and assumptions and a realistic assessment of capabilities.

MULTI-HAZARD, ALL-HAZARD, FUNCTIONAL PLANNING

There are numerous emergency management requirements that are common to any disaster situation regardless of the cause. Experience has shown that plans developed for one type of hazard can be very useful in coping with other emergency situations. Emergency management capabilities can be developed by building a foundation of broadly applicable functional capabilities in such areas as command and control, warning, communications, evacuation, and sheltering. Multi-hazard functional EOPs, therefore, begin by providing for basic capabilities without reference to any particular hazard. Hazard-specific planning, within the multi-hazard planning process, focuses on those requirements that are truly unique and not properly covered by the planned generic capability.

HISTORY OF THE ICS

The identification of a need for an ICS stems from a series of fires and multi-agency responses in the 1970s. In Southern California, a series of wildland fires over a thirteen-day period burned more than 500,000 acres in seven counties, destroying 800 structures and killing sixteen people. The economic loss approached $233 million. An analysis of the emergency response to these fires found no management mechanism or resource allocation process in place that could coordinate the federal, state, and local resources needed to respond effectively to future wildland fire emergencies.

In 1972, Firefighting Resources of Southern California Organized for Potential Emergencies (FIRESCOPE) was established to assist the fire services in Southern California with the coordination of multi-agency emergencies or disasters that could not be contained by one jurisdiction's resources. Congress provided the United States Forest Service's Fire Research Laboratory with funds to assist in the development of this coordinated response plan.

FIRESCOPE developed what is known today as the ICS. It is designed to be used as a multi-agency, multi-jurisdiction, all-hazard system of command and control. Although originally designed for responses to wildland fires in Southern California, the ICS is now used throughout California by most emergency response agencies in almost all disaster situations. The ICS is designed to include all agencies that may be involved in emergencies or disasters, such as fire services, emergency medical services, law enforcement, emergency management, public works, etc.

During the early 1980s, fire departments across the country saw the success of the ICS during emergencies and disasters, and identified the need for some type of command and control system for everyday fireground use. Most fire departments used the ICS as a model, and began developing their own incident management systems for day-to-day emergency incidents, rather than for multi-agency, multi-jurisdiction use. The most notable of these systems was the Fireground Command System, developed by the Phoenix Fire Department. A very popular training package for the Fireground Command System was developed and marketed to fire departments nationwide by the National Fire Protection Association (NFPA).

In 1987, shortly after the passage of NFPA 1500 (which is the standard for the Fire Department Occupational Safety and Health Program), the NFPA Technical Committee on Fire Service Occupational Safety and Health initiated a project to look at the different types of incident command and incident management systems being utilized across the country. The intent of the project was to identify the main components of successful systems and to develop a generic standard to ensure that such systems would have certain common characteristics. That way, there would be some level of compatibility among systems that met this standard, regardless of which system the department used.

Key players from FIRESCOPE, the Phoenix Fire Department, and many other organizations, agencies, and departments met several times during the development of the standard. The group found more than differences in the systems, however the procedures within the systems were also different in nature for a "routine" emergency than for a large, disaster-type incident. The group agreed that each of these systems was designed to help manage an incident and thus should generically be called "incident management systems." This would also eliminate confusion between FIRESCOPE's ICS and a department's incident management system used for routine emergencies.

In 1991, following the adoption of the NFPA 1561 Standard on Fire Department Incident Management Systems, many of the groups involved in the development of the standard recognized the need to continue their efforts, working toward a single, national incident management system. Representatives of these groups formed the National Fire Services Incident Management System Consortium, which now includes representatives of about forty fire service agencies across the country.

One of the initial efforts of the consortium was to integrate the strategic and tactical procedures of the Fireground Command System with the overall structure of ICS. They have developed the *National Fire Service Incident Management System Model Procedures Guide for Structural Firefighting*. This model procedures guide, published by Fire Protection Publications of Oklahoma State University, is intended for use by fire services for incidents in which fewer than twenty-five units have been committed to the scene. The consortium recognized that ICS is the foundation for any model incident management procedures, and that beyond twenty-five units, an incident has escalated into an area best managed by FIRESCOPE's ICS.

The consortium, including representatives from FIRESCOPE and the Phoenix Fire Department, has requested that FEMA's United States Fire Administration (USFA) include the model procedures guide in the instructional curriculum of USFA's National Fire Academy (NFA) for structural firefighting operations. FEMA selected ICS as a model system in the early 1980s. ICS currently serves as the standard instructional underpinning of the curriculum offered by the NFA. The model procedures guide would be a natural extension of NFA's program and would enhance their curriculum. The consortium has recently agreed to turn over its model procedures to NFA for instructional purposes.

ICS can be used as a framework for managing hazardous materials emergencies, emergency medical service and mass casualty incidents, and other all-hazard situations. More and more organizations, including the Integrated Emergency Management System National Advisory Committee, have endorsed the ICS concept and the use of the consortium's model procedures.

EXERCISING

After the planning process has been completed, the testing operations must begin. Exercising provides an organized approach and a controlled environment for plan evaluation.

Why conduct exercises? Because of their structured and controlled approach, exercises can provide better evaluation of the EOP than actual events, in which results cannot be attributed directly to the EOP because they may be the product of chance or luck. Under Title III of the Superfund Amendments and Reauthorization Act (SARA), exercises are mandated for the testing of hazardous materials contingency planning. FEMA Region VI has one of the best exercising programs in the nation.

The variety of needs for exercising has produced a selection of exercises, each designed to suit specific circumstances. The five types of exercise activities include

- Orientation seminars
- Drills
- Tabletop exercises
- Functional exercises
- Full-scale exercises

The ongoing process for an effective emergency management system is a plan maintenance program. The program defines the circumstances that warrant upgrading of the plan and the appropriate strategies for maintaining a plan, including

- Post-emergency response evaluation
- Exercise evaluation
- Resource base changes
- New hazards
- Identifying appropriate strategies for maintaining the EOP

Maintaining the EOP is part is of the cyclical planning process. The maintenance of emergency management systems is never ending.

SUMMARY

The CEM and IEMS steps described in this chapter are intended to serve management at each level of government by providing basic information to make reasonable and justifiable plans and to take effective action to increase emergency management capabilities. Government agencies will realize benefits from the process almost immediately. It takes time, however, to achieve total integration of emergency management activities and to develop the capabilities required to perform the functions necessary to deal effectively with all hazards.

It will also take time and practical experience to refine the process and to develop the best guidance to assist in its implementation. Through cooperation and constructive criticism from emergency managers and professionals, public and private, at all levels, you will continue to make progress toward your goals. Communication center personnel should understand their importance in emergency management and be able to function under the common system used by the community at large.

Just as there is no typical emergency or disaster, there is no single emergency management organizational structure that is ideal for every community. The CEM and IEMS concepts, however, do recognize that disasters have common phases and that a systematic emergency management approach is best for dealing with these phases.

QUESTIONS

1. What is emergency management?

2. How will an effective emergency management system benefit the community?

3. What are the threats facing the United States?

4. List five potential hazards facing our communities.

5. Define the role of the emergency manager.

6. Define the role of the communications center director.

7. What is CEM?

8. Define IEMS.

9. List five resources an effective emergency management system should be able to provide.

10. What is an EOP?

CHAPTER 3

Event Planning and Management

OVERVIEW

In this chapter we will describe the basic concepts of the Incident Command System (ICS) and other command systems used during a terrorism event. It is important that communications personnel understand the standard system under which all major situations will be handled.

This chapter will enable you to

- Define crisis and consequence management and identify the federal agencies responsible for these management roles

- Identify the five functions of the ICS

- Identify the importance of a unified command during a Weapons of Mass Destruction (WMD) event

- Describe the role of the local emergency operations center (EOC) and the interface between the field Incident Command (IC) and the EOC

In assisting other jurisdictions with terrorism planning, several issues always seem to be apparent. The most common issue is that of turf and unfamiliarity with the other agency. It is interesting to hear the opinions one agency may have of another even though the two have rarely interacted with each other. The opinions seem to be formed second-hand based on other agencies that deal with them, yet the reality never seems to match the opinions formed. When multiple agencies get together, the lack of interagency understanding can lead to turf battles.

One great way to start the planning process is with a brief overview from each agency. This could include the agency mission statement, roles and responsibilities, and the available resources for radio interoperability. All agencies will then have an equal understanding of each other. If an emergency manager is in the middle of the planning process and it is not functioning well, going back to the beginning may help immensely. In some cases a particular agency may overestimate its own role in an emergency. Departments need to look at their responsibilities and learn to give up some control as needed during emergency situations. During large events, state and federal resources may overwhelm local personnel. Because communications centers are behind the scenes, they are often unappreciated by the community and media. Other agencies may feel communications centers report to and work for them, when in fact communications centers have a unique role in emergency response planning as a stand-alone division or agency. communications centers need to know which agencies will respond as mutual aid and what radio/communications systems they will require. The overall goal is to learn to work together at the local, state, and federal level during a terrorist event. The following material covers coordination of multiple agencies.

One goal of emergency management is to coordinate a unified plan for response to a crisis using the CEM and IEMS ideas. The terrorism/emergency managers and planners are not expected to be an expert in every area, but will be expected to know where to get expert help. Communications should work closely with emergency management to make sure that what you *can* provide is planned for and to make sure there are no unrealistic expectations of the communications system. The emergency management office is a broker for resources and should have a list of experts, equipment, and supplies for disasters.

CRISIS AND CONSEQUENCE MANAGEMENT

According to Presidential Decision Directive 39 (PDD-39), the FBI is the federal agency responsible for the management of the crisis created by an act of terrorism. Initially, the local law enforcement agency will assume this responsibility, but it will pass authority to the FBI when they arrive on the scene. Management of the consequences of the act (i.e., building collapse, search and rescue, medical treatment, etc.) begins with the local responders. It is supplemented by the Federal Emergency Management Agency (FEMA), which is able to provide many resources that may be beyond the reach of the local community or even the state.

Local first responders are going to be saddled with the initial brunt of the event. Just as a building needs a foundation that is engineered to hold the building, the local community must also provide the foundation for the multitude of resources that will soon arrive. The quicker a management system is developed—not just at the scene but throughout the entire community—the better the event can be managed. Having an incident management system in place is not an option. It is a necessity.

There are different versions of incident management systems throughout the country—in the military and in local, state, and federal governments. The concepts presented in this book are somewhat generic. The goal of this section is to show how these different incident management systems can work together using their commonalities to fit them into one operating process.

The resources and agencies that will assist local agencies expect some sort of management system to be in place and operating when they arrive. Failure to provide the framework for the outside resources to plug into will jeopardize responder safety, result in unproductive actions, and may be measured in terms of lives lost or injuries.

The intent of this discussion is to present examples of management systems and specific organizational structures so that the commonalities can be observed and capitalized upon to ensure the best response and recovery operations possible.

INCIDENT OPERATIONS

Most Incident Management Systems (IMS) or ICSs consist of five sections: command, operations, planning, logistics, and administration or finance. Each of these sections is responsible for certain functions of the incident. With a unified command function directing regularly scheduled planning and briefing cycles, many agencies can work cooperatively to bring the incident to a safe conclusion.

A unified command structure should be implemented whenever multiple agencies have jurisdiction in an event, such as when the event crosses jurisdictional boundaries (e.g., floods covering numerous communities) or when multiple agencies have legal responsibility in an event (e.g., hazardous materials events with involvement of environmental regulators, Emergency Medical System (EMS), and fire services). Unified command principles ensure that objectives are developed that meet the needs of all the participating agencies. Planning meetings with all agencies represented will allow a comprehensive Incident Action Plan (IAP) to be developed for each operational period. This IAP, used in shift briefings, helps ensure consistency of efforts and results in safer operations.

Local responders should practice the principles of an IMS on every call. The interaction and responsibilities of fire services, law enforcement, and EMS on a scene should not be an issue. What matters are the cooperative efforts of each agency that will allow each of them to complete their responsibilities during the event.

A typical organizational chart, built by local responders for a reinforced and extended operation, might look like those shown in Figure 3-1 and Figure 3-2.

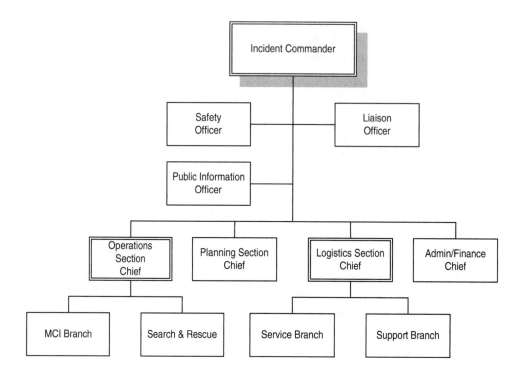

Figure 3–1 Local Responder organizational chart.

EOC OPERATIONS

The next level of management and coordination within a community will most likely occur in or at the EOC. Administrative functions are usually given to the EOC staff and the incident IC retains on-scene authority. The EOC manages the entire community during the incident, and the IC manages the actual incident. These functions may sometimes overlap and conflicts may develop. In these situations, the EOC, which has responsibility for the entire community, will prevail. The EOC will not dictate actions at the scene, but may influence changes in the action plan or goals by advising on resource availability. While the on-scene IC is making plans and directing operations at the event site, the IC's resource requests will most likely be directed to the EOC when it is activated. The EOC is the point from which requests are made for state resources (Figure 3-3).

Most state EOCs have positions called Emergency Support Functions (ESFs). Federal resources are managed through the ESF categories and federal coordinating centers are organized around the ESF concept. For example, ESF-4 is Fire Fighting and ESF-8 is Health and Medical Services. On a federal level, the United States Public Health Ser-

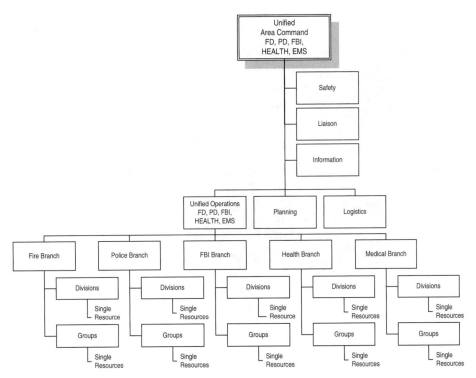

Figure 3–2 Unified Command System.

vice (USPHS) is the lead agency for ESF-8. An example of a federal resource that can assist through ESF-8 is a Metropolitan Medical Strike Team (MMRS).

Without carefully preparing and exercising a well-rehearsed plan, confusion will reign as local responders try to communicate through their EOC to resources that understand the ESF language. The meshing of an IMS and ESFs usually takes place at the local EOC. This is referred to as the ICS/EOC interface.

If local governments expect to be able to fully utilize the state and federal resources available to them, they need to work toward making the ICS/EOC interface a smooth transition from scene actions to support. Again, outside resources are looking for a place to plug in to efforts already underway and to assure of the safety of their personnel.

INTEGRATION WITH FEDERAL GOVERNMENT

Local ICs and leadership must recognize that the FBI has full authority (through PDD-39) over Nuclear, Biological, or Chemical (NBC) terrorist events, but will operate in a unified command structure with the IC during the response and rescue phase. Once all viable victims have been removed, primary control will shift to the

Figure 3–3 Your local Emergency Operations Center (EOC) will be a focal point for any large disaster/event in your community.

FBI Special Agent in Charge (SAIC) and the local responders will operate in support of the FBI.

Under the Federal Response Plan, federal agencies are responsible for providing coordinated assistance to supplement state and local resources in response to public health and medical needs following a significant natural disaster or man-made event when federal assistance is requested. This support is categorized into the following areas:

- Assessment of health/medical needs
- Health surveillance
- Medical care personnel
- Health/medical equipment and supplies
- Patient evacuation
- In-hospital care
- Food/drug/medical device safety
- Worker health/safety
- Radiological hazards
- Biological hazards
- Chemical hazards

- Mental health

- Public health information

- Vector control

- Potable water/waste water and solid waste disposal

- Victim identification/mortuary services

Resources available from the federal government include the National Disaster Medical System (NDMS). This resource includes Disaster Medical Assistance Teams (DMATs), which can be mobilized for a terrorist event. At the end of this chapter, a list of federal resources (by agency) are listed. The list is in abbreviated form, so check with each agency for additional planning information (Figure 3-4).

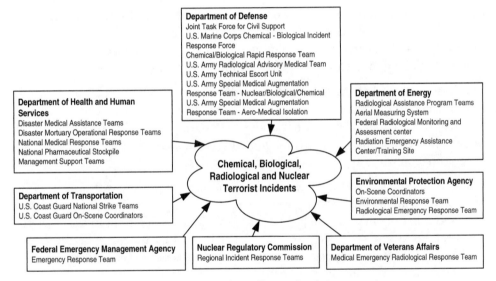

Figure 3–4 Federal agencies that will assist locals in a terrorist event.

The Department of Defense's (DoD) role is articulated primarily through the Federal Response Plan (FRP) in ESF-3: Corps of Engineers and ESF-9: Urban search and rescue. There are, however, specialized agencies within the DoD that could serve usefully in nuclear, biological or chemical (NBC) events. These include the Technical Escort Unit (TEU) in Aberdeen, Maryland, the Chemical/Biological Incident Response Force (CBIRF) in Camp Lejeune, North Carolina, and the antiterrorism team attached to the Edgewood Research, Development, and Engineering Center (ERDEC) located in Edgewood, Maryland. These teams can only be activated by another federal agency such as the FBI or USPHS. Because the teams' response time is likely to exceed three hours, the determination to access them must be made early in the event.

FEMA assumes federal consequence management and public safety responsibility for NBC events once the attorney general has determined that the priority law enforcement goals and objectives have been set or are outweighed by the consequence management concerns. As the primary agency for ESF-5: Information and Planning, FEMA coordinates the acquisition of federal resources for incident mitigation and activate urban search and rescue when indicated.

The Department of Energy (DOE) can play a critical role in providing specialized technical support in a nuclear terrorist event. This support may be more appropriate in a long-term scenario for agent/material removal and disposal.

The Centers for Disease Control and Prevention (CDC) are an immediate resource that should be notified as early in the incident as possible, however it is not likely that CDC personnel can be transported to the site in a timely fashion. Therefore, a reliable communications linkage should be established for the rapid exchange of information and medical consultation. The CDC can provide consultation on chemical antidotes, chemical decontamination practices, and medical intervention actions (both long- and short-term) for chemical and biological poisonings.

The Environmental Protection Agency (EPA) is the primary response agency for ESF-10: Hazardous Materials. Its role is to provide a coordinated federal response to actual or potential release of hazardous materials. In an NBC scenario, its role would involve the long-term remediation and decontamination of the incident site in coordination with other federal and state agencies.

The United States Secret Service, as a law enforcement agency with responsibility for protecting the United States government's leadership as well as visiting heads of state and other dignitaries, would have little role in an NBC event unless it jeopardized the safety and well being of these officials. At that time, the Secret Service would focus its efforts on personnel removal and protection—very little of its effort would be expended in incident mitigation.

In any scenario, it must be recognized that federal agencies and resources will not likely be activated and mobilized prior to the critical elements of an NBC event, which must be addressed by local responders. Only if the event length exceeds the twenty-four to thirty-six hour time frame will federal agencies arrive at the Incident Command Post (ICP) to provide support. Most federal assets can be accessed through the SAIC or USPHS on-site representative.

The governor of a state can activate the Army National Guard (ANG). However, this requires that a state of emergency be declared. The ANG could be useful by activating its assets to support a local response. This resource is most valuable for security and manpower.

EMERGENCY OPERATION PLAN (EOP) COMPONENTS

There are many ways to organize an EOP. Many plans follow the FEMA Community Preparedness SLG 101, formally called CPG 1-8. Its outline contains headings for the following:

- Basic plan
- Annexes
- Appendices
- Standard operating procedures

TECHNICAL WRITING

Technical writing conveys specific information about a technical subject to a specific audience. Technical writing is the science of communicating the plan to the user. The user may be a state governor or volunteer relief worker. Research who your audience is and make the EOP readable for it.

The ability to compose a plan uses technical writing skills. To illustrate how technical writing can convey information quickly consider the following experiment. During class, a United States Air Force technical manual with over 500 pages was given to a student. He was asked a technical question, and within three minutes he had the answer. The secret is to create a good table of contents, index, and logical subject progression so that the answer to any question can be found with little difficulty.

Technical writing should be or convey

- Clarity—single meaning
- Accuracy—No mistakes
- Comprehensiveness—Provides needed information
- Accessibility—Ease of locating information
- Conciseness—Shorter the better

Your local EOP should

- Address all functions
- Contain a basic plan, functional annexes, and hazard specific appendices (for requirements unique to certain hazards)
- Be updated regularly

WHAT MAKES A GOOD PLAN

A good plan

- Is based on facts or valid assumptions
- Is based on an inventory of community resources

- Provides organizational structure

- Uses simple language

- Uses coordinated elements

- Is a "living document"

- Is the first thing you reach for when an incident occurs

- Keeps concepts simple with few details

- Assigns responsibility to those who know the mission best

SIGNS OF A BAD PLAN

A bad plan

- Is used as door stop

- Weighs over three pounds

- Has dust on it

- Cannot be found or is absent during a disaster

- Only one copy is available in agency

- Has outdated phone numbers and names

TESTING THE EOP

For assistance in designing an exercise, see FEMA's exercise design course.

Types of Exercises

- Orientation

- Table top

- Functional

- Full-scale

- Actual event (can count as exercise)

ESFs

Although many communities have developed functional annexes which are unique to the locality, a general starting point for identifying which annexes will be required can be obtained by looking at state and federal response plans.[1] Generally speaking, there are twelve basic annexes that were identified when we discuss the Federal Response Plan.

The Federal Response Plan adopted a new format utilizing ESFs. These are the common functional areas similar to the functions (direction and control, medical, transportation, warning) outlined in CPG 1-8 or SLG-101. The ESFs are

---

1. For a copy of the Federal Response Plan, see *www.fema.gov* and type "FRP" in the search engine.

ESF 1: Transportation

Providing civilian and military transport
Lead agency: Department of Transportation

ESF 2: Communications

Providing telecommunications support
Lead agency: National Communications System

ESF 3: Public Works and Engineering

Restoring essential public services and facilities
Lead agencies: United States Army Corps of Engineers, DoD

ESF 4: Fire Fighting

Detecting and suppressing wildland, rural, and urban fires
Lead agencies: United States Forest Service, Department of Agriculture

ESF 5: Information and Planning

Collecting, analyzing, and disseminating critical information to facilitate the overall
federal response and recovery operations
Lead agency: FEMA

ESF 6: Mass Care

Managing and coordinating food, shelter, and first aid for victims; providing bulk
distribution of relief supplies; operating a system to assist family reunification
Lead agency: American Red Cross

ESF 7: Resource Support

Providing equipment, materials, supplies, and personnel to federal entities during
response operations
Lead agency: General Services Administration

ESF 8: Health and Medical Services

Providing assistance for public health and medical care needs
Lead agencies: United States Public Health Service (USPHS), Department of Health
and Human Services

ESF 9: Urban Search and Rescue

Locating, extricating, and providing initial medical treatment to victims trapped in
collapsed structures
Lead agency: FEMA

ESF 10: Hazardous Materials

Supporting federal response to actual or potential releases of oil and hazardous
materials
Lead agency: EPA

ESF 11: Food

Identifying food needs; ensuring that food gets to areas affected by disaster
Lead agencies: Food and Nutrition Service, Department of Agriculture

ESF 12: Energy

Restoring power systems and fuel supplies
Lead agency: Department of Energy (DOE)[1]

The following section contains points to address and consider when developing a plan or revising your existing plan. The lists that follow are suggestions for a terrorism and biological terrorism plan. You can refer to these ideas; however, do not limit yourself to the following list. THINK OUTSIDE OF THE BOX—THEY DO!!!!!!!

In addition to the primary ESF functions, additional annexes may be required depending upon your specific jurisdiction. Such additional annexes may include law enforcement and security, volunteers/donations, public information/warning, and military support.

The ESFs are responsible for coordinating the acquisition and deployment of resources as needs are identified. Each ESF has one primary agency and then is supported by additional agencies. ESF-1, Transportation, for example, may be served by the county's transportation department and supported by the local school district's transportation department, as well as by utilities and the road and bridge department, etc. This group of agencies then coordinates the acquisition of resources as requested by field units or the overall commander (normally the county manager, elected officials, or their designees).

To provide greater coordination of the ESFs, they may be further subdivided along incident command functions. In our model community, the ESFs are divided along ICS lines, as depicted in Figure 3-5.

By utilizing support functions that have been established within the local communities (that is, the EOP), the resources of not only your EMS agency but all governmental agencies within your community can be accessed to assist in mass gathering events, major emergencies, or disasters. Furthermore, once the incident exceeds the local community's capabilities, the local community declares a state of emergency and the state EOP is activated (bringing into play the resources and agreements that were discussed in earlier units). Once the state resources have been exceeded, the governor of the state can request a federal declaration of a major emergency from the President of the United States, which then activates the Federal Response Plan.

The enormous resources that can be developed under the local EOP make it an invaluable tool to the EMS system. In order to ensure the effective and efficient

---

1. This reference can be found on FEMA's website, *www.fema.gov.*

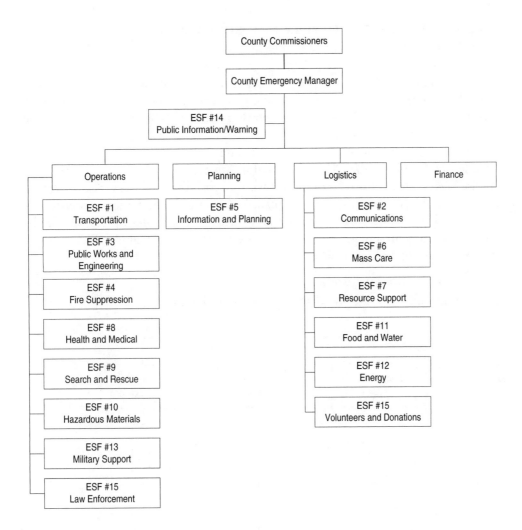

Figure 3–5 Incident Command System as it falls under the ESF functions.

operations of the plan, however, it must be utilized and tested on a regular basis. Therefore, any possible opportunity to implement the system should be seized rather than wait for a disaster or major emergency to test its operation. Many communities implement the plan regularly during special events and mass gatherings to test and exercise the plan and its participants. Limited plan activation can help identify weakness, build upon strong points, and provide an excellent training opportunity for the support function personnel.

INTEGRATION WITH STATE/LOCAL GOVERNMENT

Once the initial response to a WMD incident has occurred and the local responders are on-scene and have requested additional resources such as Urban Search and Rescue (USAR) or MMRS etc., state emergency service agencies will be notified by and through the local 9-1-1 center or local emergency management agency. EOCs may decide to provide optimal support and coordination for the locality affected by the incident. The state hazardous materials officer or other appropriate state representative should report to the ICP to coordinate state activities in concert with the task force leader. Similarly, the MMRS law enforcement sector should coordinate with on-site law enforcement to accomplish those tasks assigned by the task force leader and identified in other sections of the plan. Finally, the MMRS medical operations sector should coordinate with state and local EMS, public health offices, and medical community representatives. The MMRS was used as an example. This example can be used for other federal agencies.

USAR and MMRS Organizational Chart

An organizational chart for MMRS is presented in Figure 3-6.

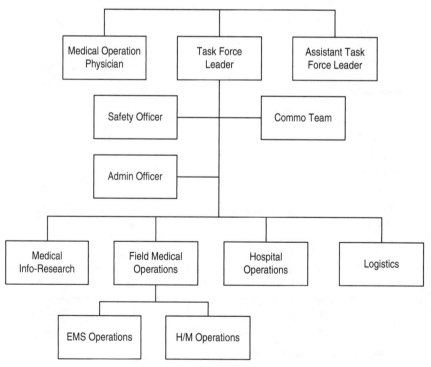

Figure 3–6 MMRS organizational chart.

An organizational chart for USAR is presented in Figure 3-7.

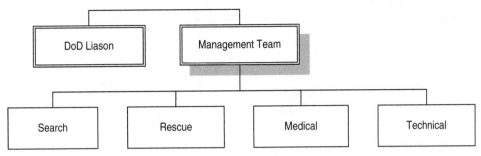

Figure 3–7 IC organizational structure.

FEDERAL RESPONSE TEAM MISSIONS AND FUNCTIONS

In the event that federal assistance is needed for a terrorist incident, telecommunicators should have a brief understanding of what federal resources are available and what they will do for their community. Table 3-1 outlines the various response teams' activity.

Table 3–1: Response Team Missions and Functions

DEPARTMENT OF DEFENSE	
Joint Task Force for Civil Support	
Mission	Function
Supports lead federal agency, establishes command and control of designated DOD forces, and provides military assistance to civil authorities to save lives, prevent human suffering, and provide temporary critical life support. Sixty dedicated personnel located at Fort Monroe, Va. Travels by military aircraft or ground transportation. Initial team deploys within 4 hours.	NBC Incident: Commands all federal military forces on site for consequence management and coordinates these activities with the lead federal agency.

Table 3–1: Response Team Missions and Functions (continued)

DEPARTMENT OF DEFENSE	
Chemical/Biological Rapid Response Team	
Mission	Function
Coordinates and integrates technical assistance for the neutralization, containment, dismantlement, and disposal of chemical or biological materials, and assists first responders in dealing with consequence management. Fourteen dedicated personnel located at Aberdeen Proving Grounds, Md. Travels by commercial or military aircraft or ground transportation. Initial team deploys within 4 hours, and remainder of team deploys in 10 to 12 hours.	Chemical/Biological Incident: Provides specialized technical advice to the Joint Task Force for Civil Support. Offers links to U.S. Army experts in a variety of disciplines, such as agent detection and disposal and assistance from medical laboratories.
U.S. Army Technical Escort Unit	
Mission	Function
Provides chemical/biological advice, assessment, sampling, detection, field verification, packaging, escort, and render safe for chemical/biological devices or hazards. One hundred ninety-three dedicated personnel located at Aberdeen Proving Grounds, Md.; Fort Belvoir, Va; Pine Bluff, Ark.; and Dugway, Utah. Travels by military aircraft or ground transportation. Team deploys in 4 hours.	Chemical Incident: Samples, detects, and identifies chemical agents. Renders safe, packages, and escorts chemical munitions or devices.
U.S. Army Special Medical Augmentation Response Team (Nuclear/Biological/Chemical)	
Mission	Function
Provides technical advice in the detection, neutralization, and containment of chemical, biological, or radiological hazardous materials in a terrorist event. Six teams located at various sites with 6 collateral duty members per team. Travels by military aircraft or ground transportation in 12 hours.	NBC Incident: Provides advice to (1) medical treatment facilities on handling contaminated patients and (2) authorities on determining follow-on medical resources, supplies, and equipment to resolve the incident.

Table 3–1: Response Team Missions and Functions (continued)

DEPARTMENT OF DEFENSE	
U.S. Army Special Medical Augmentation Response Team (Aero-Medical Isolation)	
Mission	Function
Provides a rapid response evacuation unit to any area of the world to transport and provide patient care under conditions of biological containment to service members or U.S. civilians exposed to certain contagious and highly dangerous diseases. Approximately 20 collateral duty personnel at Fort Detrick, Md. Travels by military aircraft.	Chemical Incident: Samples, detects, and identifies chemical agents. Performs casualty search, extraction, and decontamination. Performs triage and emergency medical treatment in a contaminated zone. Performs first aid, advances cardiac life support, and trauma support initially for 250 patients. Administers 1,500 nerve agent antidotes. Biological Incident: Provides highly specialized patient care during evacuation to medical facilities. Provides limited patient care in isolation units.
U.S. Marine Corps Chemical-Biological Incident Response Force	
Mission	Function
Provides force protection or mitigation in the event of a terrorist incident, domestically or overseas. Three hundred seventy-three dedicated personnel at Indian Head, Md. Travels by military aircraft or ground transportation. Initial team deploys in 6 hours, and remainder of team deploys in 24 hours.	Biological Incident: Has capability to detect four biological agents. Chemical Incident: Samples, detects, and identifies chemical agents. Performs casualty search, extraction, and decontamination. Performs triage and emergency medical treatment in a contaminated zone. Performs first aid, advanced cardiac life support, and trauma support initially for 250 patients. Administers 1,500 nerve agent antidotes.
U.S. Army Radiological Advisory Medical Team	
Mission	Function
Assists and furnishes radiological health hazard guidance to the on-scene commander or other responsible officials at an incident site and the installation medical authority. Eight to 10 collateral duty personnel located at Walter Reed Army Hospital, Washington, D.C. Travels by military transportation, commercial aircraft, or personal vehicles within 8 hours.	Nuclear Incident: Monitors contaminated medical facilities and equipment. Provides guidance about health hazards from radiological contamination. Provides advice for appropriate medical treatment. Can treat victims.

DEPARTMENT OF HEALTH AND HUMAN SERVICES

Management Support Teams

Mission	Function
Manage federal medical teams and assets that are deployed in response to an incident. Six to 8 dedicated personnel located at Rockville, Md., supplemented by 18 to 20 collateral duty Department of Veterans Affairs personnel. Travel by commercial or military aircraft. Initial team (2 to 5 members) expected to be ready to deploy within 2 hours and arrive within 12 hours. Full team expected to arrive within 12 to 24 hours.	N/B/C Incident: Coordinate the activities of federal civilian medical teams.

National Medical Response Teams

Mission	Function
Decontaminate casualties resulting from a hazardous materials incident, provide medical cars, and deploy with pharmaceutical cache of antidotes and medical equipment. Four teams located at Washington, D.C. (non-deployable); Winston-Salem, N.C.; Denver, Colo.; and Los Angeles, Calif., with 36 collateral duty members per team. Travel by commercial or military aircraft or ground transportation. Expected to be ready to deploy within 3 hours and arrive within 12 hours.	Chemical/Biological Incident: Collect and secure contaminated material, e.g. victims' clothing and any items that are circumspect after initial search for transition to crisis management responders. Provide extensive decontamination capability. Perform casualty triage. Provide extensive medical care, stabilize patients, and administer antidotes and other medications. The teams each have a supply of pharmaceuticals to treat 5,000 people. Nuclear Incident: Extract and decontaminate victims from contaminated area. Perform casualty triage. Provide limited medical care.

Disaster Medical Assistance Teams

Mission	Function
Provide emergency medical care during a disaster or other event. Forty-four teams at various locations nationwide with 34 collateral duty members per team. Travel by commercial or military aircraft or ground transportation. Expected to be ready to deploy within 3 to 4 hours and arrive within 12 to 24 hours.	NBC Incident: Perform casualty triage. Provide emergency medical care and patient stabilization. Assist in the transport of victims from medical points of distribution to medical facilities such as area hospitals.

DEPARTMENT OF HEALTH AND HUMAN SERVICES	
Disaster Mortuary Operational Response Teams	
Mission	Function
Provide identification and mortuary services to state and local health officials upon request in the event of major disasters and emergencies. Ten teams at various locations nationwide with 34 collateral duty members per team. Travel by commercial or military aircraft or ground transportation. Expected to be ready to deploy within 4 hours and at the site within 6 to 12 hours	N/B/C Incident: Perform recovery, identification, and processing of fatalities. Provide advice on the effects of decomposing fatalities. Decontaminate fatalities.
National Pharmaceutical Stockpile	
Mission	Function
Resupplies state and local public health agencies with pharmaceuticals and other medical treatments in the event of a terrorist incident. Four to 6 dedicated personnel located at Atlanta, Ga. Travels by commercial, charter, or military aircraft. Expected to arrive within 12 hours.	Biological Incident: Personnel who accompany the stockpile advise and assist in the organization of bulk stockpile medications into individual doses. They also advise and assist in the implementation of plans to distribute and dispense stockpile medications.

DEPARTMENT OF ENERGY	
Radiological Assistance Program Teams	
Mission	Function
Assist federal agencies, state and local governments, private business, or individuals in incidents involving radiological materials. Twenty-six teams at various locations nationwide with 7 collateral duty members per team. Normally travel by ground transportation but can deploy by commercial aircraft. Expected to arrive within 2 to 6 hours.	Nuclear Incident: Conduct initial site assessments. Small, regionally based teams provide quick response capability to calls for radiological assistance. Advise decision makers on steps that can be taken to evaluate and minimize the hazards of a radiological emergency.

DEPARTMENT OF ENERGY

Federal Radiological Monitoring and Assessment Center

Mission	Function
Collects, evaluates, interprets, and distributes off-site radiological data in support of the lead federal agency, state and local governments. Coordinates federal resources in responding to the off-site monitoring and assessment needs at the scene of a radiological emergency. Team members deploy in phases. Phases I (15 members) and II (45 members) consist of collateral duty. DOE personnel from Nellis Air force Base, Nev., and other locations. Phase III (known as full Federal Radiological Monitoring and Assessment Center) involves multiple federal agencies and may have 150 or more personnel from various federal agencies. Travels by military, commercial, or DOE-owned aircraft. Expected to arrive within 4 to 8 hours (Phase I), 11 hours (Phase II), and 24 to 36 hours (Phase III).	Nuclear Incident: Acts as the control point for all federal assets that are monitoring and assessing off-site radiological conditions. Gathers and assesses radiological data from multiple sources, including Radiological Assistance Program teams and the Aerial Measuring System. Also provides assessments to the state and the lead federal agency.

Aerial Measuring System

Mission	Function
Detects, measures, and tracks ground and airborne radioactivity over large areas using fixed-wing and rotary-wing aircraft. Five to 10 dedicated and collateral duty personnel located at Nellis Air force Base, Nev., and Andrews Air Force Base, Md. Initial team travels in fixed-wing aircraft and is expected to arrive within 4 to 8 hours.	Nuclear Incident: Detects and surveys the location of radioactive material deposited on the ground or the path of a radioactive plume. Fixed-wing aircraft provide quick surveys over a large area to determine the severity of the incident. Rotary-wing aircraft provide more detailed measurements.

Radiation Emergency Assistance Center/Training Site

Mission	Function
Provides medical advice and on-site assistance in triage, diagnosis, and treatment of all types of radiation exposure events. Four to 8 dedicated personnel located at Oak Ridge, Tenn. Travels by commercial or charter aircraft. Expected to be ready to deploy within 4 hours.	Nuclear Incident: Provides medical consultation and on-site assistance for the treatment of all types of radiation exposure incidents.

DEPARTMENT OF TRANSPORTATION

U.S. Coast Guard National Strike Teams

Mission	Function
Respond to all oil and hazardous substance pollution incidents in and around waterways to protect public health and the environment. Area of responsibility includes all Coast Guard Districts and Federal Response Regions. Support Environmental Protection Agency's On-Scene Coordinators for inland area incidents. Three teams located at Fort Dix, N.J.; Mobile, Ala.; and Novato, Calif., with 35 to 39 dedicated members per team. Travel by military aircraft or ground transportation. Expected to deploy within 1 to 6 hours and arrive within 12 hours.	Chemical Incident: Identify environmental contamination of waterways. Decontaminate, collect, and secure contaminated material in waterways.

U.S. Coast Guard On-Scene Coordinators

Mission	Function
Coordinate all containment, removal and disposal efforts during a hazardous release incident in coastal or major navigational waterways. Approximately 50 dedicated personnel in pre-designated Coast Guard regional zones at various locations nationwide. Travel by ground transportation. On call 24 hours. Response time depends on location of incident site.	Chemical Incident: Coordinate federal containment, removal, and disposal efforts in and around coastal waterways. Conduct initial site assessment, to include evaluating the size and nature of the released substance and its potential hazards. Direct efforts to decontaminate and clean up the incident site. Activities can include control and stabilization of the agent, on-site treatment, and off-site disposal.

DEPARTMENT OF VETERANS AFFAIRS

Medical Emergency Radiological Response Team

Mission	Function
Provides technical advice, radiological monitoring, decontamination expertise, and medical care as a supplement to an institutional health care provider. Twenty-one to 23 collateral duty personnel are located at various sites nationwide. Travels by commercial aircraft. Expected to be ready to deploy within 6 hours and arrive within 12 to 24 hours.	Nuclear Incident: Monitors for radioactivity beyond the contaminated site. Provides capability to decontaminate victims. Provides specialized medical care for radiation trauma.

ENVIRONMENTAL PROTECTION AGENCY

On-Scene Coordinators

Mission	Function
Direct response efforts and coordinate all other efforts at the scene of a hazardous materials discharge or release. Approximately 200 dedicated personnel, plus contractor support, at various locations nationwide. Travel by commercial aircraft or ground transportation Coordinators and contractors are on call 24 hours. Response time depends on location of incident site.	Nuclear Incident: Coordinate federal containment, removal, and disposal efforts. Conduct initial site assessment, to include evaluating the size and nature of the released substance and its potential hazards. Direct efforts to clean up the incident site. Activities can include control and stabilization of the agent, on-site treatment, and off-site disposal. Chemical Incident: Coordinate federal containment, removal, and disposal efforts. Conduct initial site assessment, to include evaluating the size and nature of the released substance and its potential hazards. Direct efforts to decontaminate and clean up the incident site. Activities can include control and stabilization of the agent, on-site treatment, and off-site disposal.

Environmental Response Team

Mission	Function
Provides technical support for assessing, managing, and disposing of hazardous waste. Twenty-two dedicated personnel, plus contractor support, located at Edison, N.J., and Cincinnati, Ohio. Travels by commercial aircraft. Advance team expected to deploy within 4 hours. Full team expected to arrive within 24 to 48 hours.	Chemical Incident: Offers specialized technical assistance in areas such as air sampling and ecological risk assessment. Offers specialized technical assistance in areas such as incineration and groundwater treatment.

Radiological Emergency Response Team

Mission	Function
Provides mobile laboratories for field analysis of samples and technical expertise in radiation monitoring, radiation health physics, and risk assessment. As many as 60 collateral duty personnel located at Las Vegas, Nev., and Montgomery, Ala. Travels by ground transportation or military air. Expected to arrive within 2 to 3 days.	Chemical Incident: Conducts sample preparation and analysis in mobile laboratories.

FEDERAL EMERGENCY MANAGEMENT AGENCY	
Emergency Response Team	
Mission	Function
Coordinates federal response and recovery activities within a state. Size is dependent on the severity and magnitude of the incident. Collateral duty team members are geographically dispersed at FEMA headquarters and 10 regional offices. Travel by commercial charter, or military aircraft, or ground transportation. Expected to arrive within 24 hours.	Biological/Chemical Incident: Establishes field office if required. Provides disaster assessment coordination and expertise to states and the FEMA regions.

NUCLEAR REGULATORY COMMISSION	
Regional Incident Response Teams	
Mission	Function
Carry out the responsibilities and functions of the lead federal agency during incidents at licensed facilities such as nuclear power plants. Four teams located in Atlanta, Ga.; Lisle, Ill.; Arlington, Tex.; and King of Prussia, Penn., with 25 to 30 collateral duty members per team. Travels by commercial or charter aircraft or ground transportation. Initial team expected to arrive within 6 to 12 hours.	Chemical Incident: Lead and coordinate federal actions related to the radiological technical response at incident site. Review actions the regulated entity is taking to correct problems. Provide analysis and consultation for actions taken to protect public health and safety.

SUMMARY

Many resources will be involved in the response to a terrorist event. Many of these resources will operate at the scene while others will be activated within the community, such as in hospitals and other medical care facilities. Resources from outside of the local jurisdiction will come to assist and fulfill their responsibilities. The local community must provide the management structure in which these resources can operate for the most favorable event outcome.

Incident management begins with the arrival of the first responders, who establish an incident management system for the scene. Many local responders will activate their emergency operations plans.

The local EOC will be activated and ESFs staffed. The state's EOC will be activated and ESFs staffed. Federal resources will come to the scene, and will assist the state's EOC and the local EOC.

It is imperative that the management structure that is developed is capable of plugging all resources into the operation.

For additional information on the USAR and MMRS and the latest updates and events, see *www.mmrs.hhs.gov.*

QUESTIONS

1. What is PDD-39?

2. List the five sections of Incident Command.

3. What is Unified Command?

4. What does the EOC manage?

5. List three things your local emergency plan should do.

6. What makes a good plan?

7. List five ESFs.

8. What is USAR?

9. What is the mission of the National Medical Response Team?

10. What is the mission of the Radiological Emergency Response Team?

CHAPTER 4

Training

OVERVIEW

This chapter will show the need for terrorism training for communications centers and outline the various ways to receive that training, whether it is from the user agency or from other local, state, and federal resources. The chapter also looks at resources that communications centers can use for information on training and questions about terrorism response. At the end of this chapter you will know

- Where communications centers can get training
- What courses are offered for terrorism training
- Where to access the full listing of federal terrorism training courses
- Where to get technical assistance for terrorism training

INTRODUCTION

The best emergency plan is only as good as the people who must use it. In this case, if the telecommunicators do not know the plan inside and out, it will not work as intended. Your emergency response plans to terrorist or WMD incidents will require the full participation of telecommunicators, often in many communications (comm) centers, coordinating actions and notifications with first responders, their agencies, and outside resources.

Look at your plans. How many components require early participation by the affected comm centers? Will your telecommunicators be the ones paging and dispatching special teams, notifying state and federal agencies, requesting mutual aid from surrounding agencies, coordinating law enforcement and fire-rescue responses, all in the initial phases of the incident? After all, the first hint of a serious incident will likely come in a 9-1-1 call.

How will your telecommunicators know what to do? How will they recognize an incident as a terrorist or WMD incident, especially if it starts as something smaller or even routine? Will they be able to handle the escalation of the event into a full-scale terrorist or WMD attack? Will they put your emergency response plans into action at the earliest signs of that escalation? Will responders be endangered in the meantime?

The only way to move from theory to reality, or, in this case, from plan to action, is through training and practice.

Make sure your plans address the role of telecommunicators. Their parts in the plan should be as specific and detailed as those of first responders, from the beginning to the end of an incident. Give them as much information as possible on the probable flow of events in the field to let them respond appropriately in the center. If they know what each person is supposed to be doing at any particular phase in a response, they will be able to anticipate the responders' needs and prompt them if something seems to have "fallen through the cracks" in the chaos that can result from major attacks or disasters.

Because no incident ever goes by the book, probably because the terrorists don't have or refuse to follow the book, no plan can cover every contingency. To bridge the resulting gaps, responders and telecommunicators will need sound judgment and good instincts. They will need a sound basis for decision-making when faced with unexpected developments.

Field responders get many opportunities to develop that basis. They are put through drills and training until they can respond without hesitation to the expected developments for any type of incident. Then they practice more, with levels of difficulty added to their training scenarios, so they learn to respond appropriately to these curve balls as they occur and make good "command" decisions.

Unfortunately, the traditional role for telecommunicators in these scenarios is practically nonexistent. Their role is perfunctory and incidental, even unnecessary to the success of a drill. In fact, if a telecommunicator is not available to participate, the drill commences without one; that part might be filled by a field unit who cannot participate fully in the field exercise, perhaps due to injury or illness.

In a typical drill that does involve the comm center, one telecommunicator might be assigned to monitor the training exercises on a designated radio channel and acknowledge radio transmissions from the field. Certainly, the entire center would not be involved; a supervisor and the selected telecommunicator might even be the only ones aware that a drill is taking place. The involved telecommunicator might have a copy of the drill and the expected responses to those radio transmissions. Generally, the telecommunicator would not be briefed with the field units before the drill and would not be included in the post mortem analysis of the drill.

In a real incident, the entire comm center would be required to respond immediately and in full force. Off-duty employees would be called in, vacations would be cancelled and days off would be suspended. The comm center would be a hub of activity and the point of contact for the participants, the media, and the public. Are your personnel ready?

WHAT TRAINING IS AVAILABLE?

There have been a great number of training programs and courses developed for first responder training, yet there are few, if any, that specifically relate to the role of 9-1-1 communications personnel. This lack of training is in part because of the role that comm centers have played in the past. Comm center policies and procedures are generally designed and set up by the user agencies. How responses are to be handled, who should be sent, how communications should be handled and what part communications personnel will play in the overall event are dictated by the first responders. But comm centers are the *first* first responders. In many cases the first notification of a WMD incident will be to the 9-1-1 comm center. Now is the perfect time to stand up as a comm center and participate in first responder training.

There are many agencies training our police, fire, and medical agencies. If comm center personnel are to be valuable team members of a WMD response, we need to know what they know, why they respond as they do, and be a second set of knowledge to help save the lives of first responders and the public.

We can do this with proper training and understanding of local, state, and federal terrorism response assets. We need to understand the role of communication in these plans. There is a new recognition of what could be used as a terrorist weapon. Communications personnel need to be trained to recognize a terrorist attack and know what critical information to gather from callers. Personnel should know the indicators, signs, and symptoms of exposure to nuclear, biological, or chemical (NBC) agents and recognize unusual trends that indicate a NBC incident. The events of September 11th, 2001 opened the eyes of many response agencies—and even those on Capitol Hill—to the larger role comm centers play in our country's first response system. We need to be ready to participate, ask for training opportunities, and become key players in terrorism preparedness, response, and recovery (Figure 4-1).

Figure 4–1 Ask to participate in training to increase operator knowledge.

HOW DO COMMUNICATIONS CENTERS GET TRAINING?

A good place to start looking for training is with your user agencies. Most first responders are already training for terrorist attacks in one form or another. Even training for large incidents or mass casualties can provide a good working knowledge for communications personnel. Once you have identified local training, decisions will need to be made for sending personnel out or having in-house training. Many agencies will come to your center to train personnel as it only benefits them to have knowledgeable coworkers. Make them aware of the role your comm center can play in the success or failure of an incident. This may not be easy, especially if you have to sway "old-timers" who think of telecommunicators as glorified secretaries. Fortunately, that attitude no longer prevails.

Once you have convinced them to involve the comm center, work with them to make sure the training is grounded in reality. You will have to commit significant time and energy to this process because many times outsiders will not have any idea what the comm center's roles are in various aspects of a disaster. You will have to teach them as you expand their training scenarios, so be prepared to answer the question "Why do we have to do that?" many times.

Training can also be provided through videos or self-study courses. In the event of strained budgets, training can be done by simply providing reading material and standard operating procedure (SOP) manuals at each console. Check with your local emergency management division about courses they can provide or where you can send your staff for training. It will be well acquainted with local, state, and federal training courses available for you to attend or be provided with on-site. There are also a large number of federal training courses available for first responders. Again, remember that comm center communicators are *first responders*. Seek out those training programs that would benefit your knowledge of terrorism response and assist you in your user agencies responses.

READING AND SELF-STUDY COURSES

If your budget does not allow for mass operator training off-site, then one approach *any* center can take is to provide reading material for increased first responder knowledge. This book is one example of an overall look at terrorism preparedness and response. Look at local, state and federal websites to identify reading material that would benefit communications personnel. Provide copies at each console and ask that it be read when time allows. This can be done on a progressive knowledge increase basis and proceed over a period of time. Each document can be initialed by personnel to make sure that everyone has reviewed the material.

"EMERGENCY RESPONSE TO TERRORISM" COURSE

There is a self-study course offered by the National Fire Academy titled "Emergency Response to Terrorism: Self Study (ERT:SS) (Q534)." This course is a self-paced, paper-based document and is designed to provide the basic awareness training to prepare first responders to respond to incidents of terrorism safely and effectively. Students who successfully complete the exam are eligible for a National Fire Academy Certificate of Training. You can request a copy of ERT:SS through USFA Publications at 1-800-238-3358 ext. 1189. It can also be ordered from the USFA website at *www.usfa.fema.gov/usfapubs*. Another self-study course available that may assist comm centers in planning and preparing for special events that may offer a target for a terrorist threat or attack is titled: "Special Events Contingency Planning for Public Safety agencies." The course covers a general overview of planning considerations and coping with special events. You can view this course at *http://training.fema.gov/EMI-WEB/is15.htm*.

EMERGENCY RESPONSE TO TERRORISM JOB AID

There is also an Emergency Response to Terrorism: Job Aid (ERT:JA) available through USFA publications. The ERT:JA was designed and produced through a joint partnership of FEMA, the United States Fire Administration (USFA), and the Department of Justice/Office of Justice Programs. The document is intended to support, not replace, the training messages of the ERT NFA curriculum. It is not a training manual but a "memory jogger" for those who have completed the appropriate level of training.

The Job Aid is divided into five primary sections that are tabbed and color coded for rapid access to information.

- Introduction (Gray)
- Operational Considerations (Yellow)
- Incident-Specific Actions (White)
- Agency-related Issues (Blue)
- Glossary (Tan)

The ERT:JA is soft plastic and sized to fit into a coverall or work jacket pocket or can be used at a communications console for reference. The Job Aid is available free of charge to fire departments ordering five or fewer copies from the USFA Publications Center. Other emergency response agencies may order one copy. Additional copies may be purchased from the US Government Printing Office (GPO). Order by title and GPO stock number 064-000-00027-6 (Phone: 866-512-1800 [toll free] and website: *www.bookstore.gpo.gov*).

NATIONAL FIRE ACADEMY

There are also courses available at the National Fire Academy (NFA) in Emmitsburg, Maryland. These courses are free of charge to emergency services personnel. The course, housing on campus and airfare are provided by the NFA. The only cost to the student is for meals. Courses relating to terrorism include

EMERGENCY RESPONSE TO TERRORISM: BASIC CONCEPTS (ERT:BC)

- (F531) Direct Delivery
- (W531) State Weekend Program

This two-day course is designed to prepare first responder personnel to take the appropriate course of action at the scene of a potential terrorist incident. The course provides students with a general understanding and recognition of terrorism, defensive considerations (biological, nuclear, incendiary, chemical, and explosive), as well as command and control issues associated with criminal incidents. When an incident occurs, the student will be able to recognize and implement self-protective measures, secure the scene, complete appropriate notifications to local, state, and federal authorities, and assist in completing a smooth transition from emergency to recovery and termination operations.

Target Audience

The primary target audience for this training includes hazardous materials, fire, and emergency medical services first responder personnel. The secondary audience includes law enforcement personnel, emergency communications personnel, jurisdiction emergency coordinators, public works managers, and public health providers.

EMERGENCY RESPONSE TO TERRORISM CURRICULUM

R817 - Emergency Response to Terrorism: Incident Management (VIP)

- (R817) Volunteer Incentive Program
- (N817) Regional Delivery
- (O817) State Training Systems Delivery

 Please note that there will be different course coding when making application.

The focus of this six-day course is on fire service response to terrorism from an incident management approach, especially in dealing with the areas of Biological, Nuclear, Incendiary, Chemical, and Explosive (B-NICE) attacks. This is an advanced-level course that presumes a working knowledge of the Incident Command System (ICS) and deals with issues such as recognizing a terrorist incident, preservation of evidence, planning and intelligence, federal response and Unified Command, hazardous materials and emergency medical services response, operations and scene control, termination, and recovery. It uses complex simulation activities as well as case studies to allow learners to apply skills and knowledge that will assist them greatly in beginning to prepare their own communities for emergency response to terrorist action.

Student Selection Criteria

This program is specifically aimed at personnel with incident command responsibility in ire services—including allied professionals such as police, military, or personnel from other government agencies who need to understand the fire service perspectives on incident management. Contact Admissions at 1-800-238-3358 for more information.

COLLEGE AND UNIVERSITY DEGREES

There are many colleges offering emergency management degrees that may help in the understanding of ICS and IEMS. There are also degrees offered for telecommunicators. One such degree offered through the APCO Virtual College is the Associate in Applied Sciences—Public Safety Telecommunicator (AAS-PST). In this degree program, there is a course on terrorism titled "Introduction to Weapons of Mass Destruction." Contact APCO at *www.APCO911.org* or Jacksonville State University at 1-800-231-JAX1 ext. 5925.

GOVERNMENT-SPONSORED TRAINING COURSES

There is also a full list of training courses in the United States Army Soldier and Biological Chemical Command (SBCCOM) publication "Compendium of Weapons of Mass Destruction Courses: Sponsored by the Federal Government" (*http://hld.sbccom.army.mil*). This compendium profiles all courses relating to terrorism training and provides a Performance Objective Matrix which outlines areas of competency and provides a quick reference to courses for specific target audiences based on objective descriptions. Table 4-1 is a quick reference source for NBC Areas of Competency. The full matrix can be found at SBCCOM's website.

Table 4–1: NBC Areas of Competency.

AREA OF COMPETENCY	DESCRIPTION
1	Know the potential use of NBC weapons.
2	Know the indicators, signs, and symptoms for exposure to NBC agents, and identify the agents from signs and symptoms (if possible).
2a	Know questions to ask caller to elicit critical information regarding an NBC incident.
2b	Recognize unusual trends that may indicate an NBC incident.
3	Understand relevant NBC response plans and SOPs and your role in them.
4	Recognize and communicate the need for additional resources during a NBC incident.
5	Make proper notification and communicate the NBC hazard.
6	Understand: (1) NBC agent terms and (2) NBC toxicology terms.
7	Know individual protection at a NBC incident (personal protective equipment).
8	Know protective measures and how to initiate actions to protect others and safeguard property in an NBC incident.
8a	Know measures of evacuation of personnel in a downwind hazard area for an NBC incident.
9	Know chemical and biological decontamination procedures for self, victims, site, equipment, and mass casualties: (1) understand and implement and (2) determine.
10	Know crime scene and evidence preservation at an NBC incident.

Table 4–1: NBC Areas of Competency. (continued)

AREA OF COMPETENCY	DESCRIPTION
10a	Know procedures and safety precautions for collecting legal evidence at an NBC incident.
11	Know federal and other support infrastructure and how to access in an NBC incident.
12	Understand the risks of operating in protective clothing when used at a NBC incident.
13	Understand emergency and first aid procedures for exposure to NBC agents and principles of triage.
14	Know how to perform hazard and risk assessment for NBC agents.
15	Understand termination/all clear procedures for an NBC incident.
16	Understand Incident Command System/Incident Management System.
17	Know how to perform NBC contamination control and containment operations, including for fatalities.
17a	Understand procedures and equipment for safe transport of contaminated items.
18	Know the classification, detection, identification, and verification of NBC materials using field survey instruments and equipment and methods for collections of solid, liquid, and gas samples.
19	Know safe patient extraction and NBC antidote administration.
20	Know patient assessment and emergency medical treatment in NBC incidents.
21	Be familiar with NBC related public health and local EMS issues.
22	Know procedures for patient transport following NBC incident.
23	Execute NBC triage and primary care.
24	Know laboratory identification and diagnosis for biological agents.
25	Have the ability to develop a site safety plan and control plan for a NBC incident.
26	Have ability to develop NBC response plan and conduct exercise of response.

The Rapid Response Information System website (*www.rris.fema.gov*) contains an abbreviated compilation of the federal training courses listed in SBCCOM's compendium. These courses have been selected for their directed focus on counter-terrorism. The courses have been organized into the following five subject areas:

NBC COUNTER-TERRORISM TRAINING

1. Chemical/Biological Countermeasures Training (CBCT)
2. NBC Domestic Preparedness Training Basic Awareness (Employee)
3. NBC Domestic Preparedness Training Incident Command Course
4. NBC Domestic Preparedness Training Responder Awareness Course
5. NBC Domestic Preparedness Training Responder Operations Course
6. NBC Domestic Preparedness Training Senior Officials Workshop
7. Preparing for and Managing the Consequences of Terrorism
8. Community Integration at a WMD Incident Site
9. Crime Scene Awareness at a WMD Incident Site
10. Federal Integration at a WMD Incident Site
11. Emergency Response to Terrorism: Basic Concepts (FEMA)
12. Emergency Response to Terrorism: Basic Concepts (DOJ)

GENERAL COUNTER-TERRORISM TRAINING

1. Consequences of Terrorism, Integrated Emergency Management Course
2. Emergency Response to Criminal/Terrorist Incidents
3. Emergency Response to Terrorism: Incident Management (FEMA)
4. Emergency Response to Terrorism: Self Study
5. Emergency Response to Terrorism: Tactical Considerations
6. Emergency Response to Terrorism: Incident Management (DOJ)

COUNTER-TERRORISM TRAINING: RADIOLOGICAL

1. Nuclear Emergency Planning
2. Fundamentals Course for Radiological Response Teams

COUNTER-TERRORISM TRAINING: CHEMICAL AND BIOLOGICAL

1. NBC Domestic Preparedness Training Technician HAZMAT Course
2. Agent Characteristics and Toxicology First Aid and Special Treatment (ACT-FAST) for Use of Auto-Injectors

3. Chemical Stockpile Agent Characteristics

4. Management of Chemical Warfare Injuries

5. Use of Auto-Injectors by Civilian Emergency Medical Personnel to Treat Civilians Exposed to Nerve Agent

MEDICAL TRAINING FOR NBC INCIDENTS

1. NBC Domestic Preparedness Training Technician—Emergency Medical Services Course

2. NBC Domestic Preparedness Training Technician—Hospital Provider Course

3. Management of Chemical Warfare Injuries

4. Use of Auto-Injectors by Civilian Emergency Medical Personnel to Treat Civilians Exposed to Nerve

VIDEOTAPE COURSE

NBC DOMESTIC PREPAREDNESS TRAINING BASIC AWARENESS (EMPLOYEE)

Course Sponsor: Department of Defense (DoD)/SBCCOM

Course Description: A video presentation designed to acquaint a diverse audience of employees (e.g., security guards, 9-1-1 operators/dispatchers, cleaning staff, ticket takers, hospital support staff, baggage handlers) at potential terrorist target facilities with the signs and symptoms associated with a NBC terrorist incident, and how to recognize and respond to such an incident. The course includes a facilitator's guide and an example 9-1-1 checklist. Length: 30 minutes.

Course Objectives: Upon completion of the training, employees should

- Know the potential for terrorist use of NBC weapons
- Be able to recognize an NBC attack
- Know how to make proper notification and communicate the NBC hazard

In addition, 9-1-1 operators/dispatchers should

- Know the questions to elicit critical NBC agent information from callers
- Recognize unusual trends that may indicate an NBC incident
- Know protective measures and how to initiate actions to protect others and safeguard property
- Know the support infrastructure and how to access it in an NBC incident

NBC Areas of Competency: 1,2,3,4,5,6,7,8,9,10

Target Audience: Civilian

Emergency Responder Group: Facility employees

Emergency Responder Levels: Awareness

Type of Instruction: Classroom TV/VCR by Gov/Contractor

Recommended Class: Limited only by facility capacity. Course is not facility-dependent.

Point of Contact (POC): Domestic Preparedness CB HelpLine

Address: United States Army Chemical and Biological Defense Command, Aberdeen Proving Grounds, Maryland 21010

Phone Number: 1-800-368-6498

Comments: This video presentation is part of the National Defense Authorization Act for FY96, Title XIV Defense Against Weapons of Mass Destruction Preparedness Training Program, which includes Senior Officials Workshop, Employee Basic Awareness Video, Responder Awareness, Responder Operations, Incident Command, Hazmat Technician, EMS Technician, and Hospital Provider Technician.

STATE AND LOCAL DOMESTIC PREPAREDNESS SUPPORT HELPLINE

BACKGROUND

The Office for Domestic Preparedness (ODP), Office of Justice Programs (OJP) is the program office within the Department of Justice (DOJ) responsible for enhancing the capacity and preparedness of state and local jurisdictions to respond to Weapons of Mass Destruction (WMD) incidents of domestic terrorism. The ODP's State and Local Domestic Preparedness Program accomplishes this through its training, exercises, equipment grants, and technical assistance programs. Assistance provided by the ODP is directed at a broad spectrum of state and local emergency responders, including firefighters, emergency medical services, emergency management agencies, law enforcement, and public officials.

In 1997, the DoD initiated the Nunn-Lugar-Domenici (NLD) Domestic Preparedness Program in response to Congressional direction, which included the creation of a chemical-biological helpline for first responders nationwide. The DOJ, through the ODP, is continuing the helpline and has added additional capabilities in response to lessons learned and feedback from local, state, and federal emergency response and support agencies.

HELPLINE OVERVIEW

The Helpline is a *non-emergency* resource available for use by emergency responders across the United States. The Helpline provides general information on all of the ODP's programs and information on the characteristics and control of WMD materials, defensive equipment, mitigation techniques, and available federal assets. The Helpline provides "customer intelligence" that will aid state and local jurisdictions in building capacity in their communities to respond to a WMD terrorism incident.

ODP PROGRAMS

The Helpline provides a unique service by which, with only one 800 number, the state and local response community and support agencies can reach all of the ODP available services, such as

- **WMD training:** Provides general information on ODP training available to state and local jurisdictions.

- **Centralized scheduling capability:** A comprehensive domestic preparedness point of contact database for scheduling NLD Domestic Preparedness Program training.

- **WMD exercises:** Provides general information on the State and Local Exercise Program and the exercise support to the cities under the NLD Domestic Preparedness Program.

- **Equipment grants:** Provides general information on the State Domestic Preparedness Equipment Program and funding availability.

- **NLD Domestic Preparedness Program:** Provides general information on services available to the NLD-designated cities for equipment, training, exercises, and technical assistance.

- **Technical assistance:** Provides general information on services available for assistance in enhancing the ability of state and local jurisdictions to develop, plan, and implement a program for WMD preparedness. The three specific areas are:

  - General technical assistance

  - State strategy technical assistance

  - Equipment technical assistance

- **Domestic Preparedness Equipment Technical Support Program:** Provides technical support to jurisdictions in the utilization, sustainment, and calibration of detection equipment. Scheduling of mobile, on-site training teams is available.

- **Domestic Preparedness Support Information Clearinghouse:** A clearinghouse of information on domestic preparedness, counterterrorism, and WMD available to state and local jurisdictions, including the ability to order publications and videos.

WMD EXPERTISE

The Helpline offers technical assistance in *non-emergency* cases to state and local emergency responders and public officials. Dedicated operators are trained to listen and respond to questions from the field pertaining to domestic preparedness issues, quickly retrieving the most current information relevant to the specific question. Skilled research staff and available subject matter experts in WMD materials and equipment provide detailed information on even the most esoteric questions. The Helpline provides information on the following subjects:

- Detection equipment

- Personal protective equipment

- Decontamination systems and methods

- Physical properties of WMD materials

- Signs and symptoms of WMD exposure

- Treatment of exposure to WMD materials

- Toxicology information

- Federal response assets

- Applicable laws and regulations

HELPLINE NUMBER AND HOURS OF OPERATIONS

The helpline number is 1-800-368-6498. The helpline is staffed weekdays from 9:00 a.m. to 6:00 p.m. EST. On weekends, holidays and after business hours, callers can leave a voice mail message.

WHAT HAPPENS AFTER TRAINING?

After you have integrated the telecommunicators into the responders' consciousness and plans, make sure you have a viable plan for the comm center. It must mesh with the responders' plans and help them flow smoothly. You will need input from every employee—from call taker to center manager—to devise as comprehensive a plan as possible.

Test the plan within your center as it develops. Look for gaps or, better yet, have all involved parties tell you what gaps they saw. Adjust and fine tune the plan as you go. This will require multiple dry runs and drills, in which you should include as much of the comm center as possible, as many times as possible. After all, you want every-one working there to know what to do, as well as when, where, how, and why.

When you are convinced your plan is as comprehensive as possible, coordinate joint drills with first responders. Do not be satisfied with playing a nominal role in their drills; make them truly joint ventures. This means your telecommunicators should go through all the motions a scenario would require in "real life."

Figure 4–2 Coordinate joint drills with your first responders.

Beyond that, add likely complications—like loss of radio towers or power—to make your center's employees feel the pressure of a true disaster. Better they should sweat it out in a drill than feel helpless in a real disaster.

Do not be left out of post-incident reviews. Your telecommunicators must know what went right, what went wrong, and why. They should not have to hear it summarized by you at a shift briefing days or weeks later. Not only do they need immediate feedback of their performances, but they also can give valuable information about the progress of an incident. If they are not included in the analysis of an incident, emergency managers and responders will be deprived of potentially life-saving information.

Do not handicap your emergency response plans by overlooking the expertise and experience of those who often are truly the first responders—albeit via telephone—to an incident. Your plans will depend on having the right people at the right place at the right time, and that is what telecommunicators do every day.

SUMMARY

There are many reasons why comm centers have been left out of the increase in terrorism training. Nevertheless, comm centers should request to be included in all training activities and drills of their user agencies. Even if the participation is in planning meetings and observation of drills, this in itself will help prepare centers to understand what their user agencies deal with and be able to effectively help in the response.

There is no reason why training cannot take place in comm centers. Simply reading material relating to terrorism or reading the emergency response plans of user agencies will help personnel do their jobs better. The closer we work with our user agencies, the better prepared we will be if or when our center must deal with a terrorist attack.

QUESTIONS

1. Why have comm centers been left out of terrorism preparedness training?

2. List three ways personnel can receive terrorism training.

3. What is the focus of the course "Emergency Response to Terrorism: Incident Management?"

4. What is SBCCOM?

5. What NBC areas of competency should communications center be able to meet?

6. What is the RRIS?

7. List three course objectives for the NBC Domestic Preparedness Training Basic Awareness.

8. What is the ODP?

9. What services does the ODP provide?

10. List five subjects with which the helpline can assist.

CHAPTER 5

Facility Security

OVERVIEW

This chapter will identify the basic issues of building security that a communications center must review. Communications personnel will understand the need for heightened security in the building and on the communications floor itself. This chapter will allow the communications center to review existing building security and revise or upgrade it to prevent or lessen the effects of a terrorist attack upon the facility. You will be able to

- Survey your communications center for security risks
- Identify building security measures
- Identify security upgrades
- Develop a bomb incident plan
- Identify remote tower security measures

FACILITY SECURITY ISSUES

Security of the communications center building is an important part of terrorism preparedness. The idea is to create a totally integrated security system, which combines personnel, technologies, and security procedures to protect property and lives. A communications center in a small community may be on a higher sense of alert, which includes questioning and looking for unknown persons in and around their facility. Vigilance to the out-of-place occurrence is increased. One possibility for greater security is to simply lock the communications center door. Some centers may share a building with other agencies. They will need to look at their security issues as well as meet with all occupants of the building to discuss overall security. A large communications center that is the only occupant of a building will have greater control over the level of security for its facility. Many questions need to be asked of a communications center and the level of security measures. Refer to the sample security checklist in Table 5-1 that covers many of the aspects of building security. Use this list as a starting point for your own security analysis and use it to create or upgrade your security systems.

COMMUNICATIONS CENTERS ARE AT RISK FOR ATTACKS

There is no doubt that as other areas of our country become hardened to terrorist attacks, the facilities that are less prepared will be more attractive targets. The easier it is to attack and interrupt daily life, the greater the risk. Imagine a terrorist attacking a large event or facility. Then imagine a secondary attack—not at the site of the first attack, which would be to injure or kill first responders—but at the communications center, which would eliminate the communications system for the responders and for the general public. Imagine wanting to report a terrorist bombing and not having any way to reach emergency communications. Remember the overwhelming use of cellular phones after a large disaster—what if there is no way to call for help?

Eighty percent of terrorists still prefer bombs and conventional explosives. There is little that you can do to protect a building from a large attack such as the one at the World Trade Center in 2001, but you can minimize injury and death with various building design and protections. A building should be built with the idea that it will be the last building standing after any event, whether natural or man-made.

The idea is to deter, detect, and limit access of vehicles and persons to buildings, communications centers, and supporting equipment access areas.

SECURITY SURVEY

The first thing to do is to develop a team of people who can institute a security survey. This team should include management, facility and maintenance personnel, communications supervisors, operators, and vendors who can speak to the specifics of the phone, power, and cable systems. The first meeting can include a review of existing policies, procedures, and protections that are in place to determine the level of your facility's security. One must always consider local building codes, evacuation requirements, and local security laws. If there is a greater concern for security, and if the budget allows it, hire a security consultant to provide the needed expertise. This committee can even visit local jails, courthouses, or airports to see the levels of equipment, security issues, access, and workability. Meet with the managers of the facility to discuss management, maintenance, and costs of the various security systems in place.

After this is complete, look at each component of your facility. Start on the outside and work to the inside, examining each area for security lapses and ways to improve. Decisions will need to be made to determine how much security will interfere with the access to the building. The greater the risk, the greater the need for security.

OUTSIDE THE FACILITY

If your facility can regulate parking, most experts agree you should prohibit parking within 100 feet of the building. This will prevent the type of car or truck bombs that caused the Murrah Federal Building destruction in 1995. Some buildings may be next to a public street or alleyway that cannot be restricted. If this the case, you will need to have heightened awareness of vehicles parked outside of your building. If the building has underground parking, the recommendation is to limit it to known tenants only. Truck access to underground loading docks should be strictly controlled with documents, and vehicles should be searched prior to entry. During security alerts or specific threats to the building, underground parking may have to be denied.

Vehicle traffic approach speeds can be regulated by installing curved roadways or speed bumps in the roadways to slow down approaching vehicles. You can control vehicle access by using barriers, traps, or hydraulic barricades. Ditches, water, and rock features can be used to deter vehicle ramming and decorative concrete barriers, planters, or knee walls can be placed around the building to prevent such ramming. The White House, public buildings, and even some convenience stores have these barriers, and in the days after September 11th, some buildings were protected by heavy equipment trucks placed end to end as a temporary shield.

Trash bins or garbage containers are places to hide a bomb; keep these 100 feet away from the building as well, and if they are on rollers, chain them to prevent the bin from being moved closer to the building. Lighting is essential, as poor lighting can assist intruders. Lighted areas should be checked regularly to ensure the lights are working effectively. Buildings should be well lit, and lighted areas should have some kind of surveillance, such as patrols or cameras. If possible, security cameras should be placed at entry points and high-risk areas. Cameras come in various types, including timers, photoelectric cells, and heat or movement detectors. Outside security patrols with radio communications can provide visible security protection.

BUILDING PERIMETER

Look at the building from the outside and identify any security concerns. An obvious concern is for window protection. Windows are often the weakest component of a building. A bomb blast, close to or even up to four or five blocks away, can injure or kill building occupants.

Blast film can be expensive, yet the protection offered is worth the price. A communications center can budget the entire amount at one time, or break it up into a few windows each year, starting from the first floor (the highest risk). The idea is not

Table 5–1: Security Survey

QUESTION	YES	NO
Have you done a risk evaluation of your location?	❏	❏
Have you located areas that are vulnerable to vandalism or forced entry?	❏	❏
Have you sought advice on security measures from the local police or security industry?	❏	❏
Have you sought professional guidance on the location, quality, and capacity of security equipment to meet risk level?	❏	❏
Do you set aside a specific amount of money each year and establish a rolling program for crime and vandal protection measures?	❏	❏
Has your staff been advised to report suspicious activities or strangers on the premises?	❏	❏
Do you have a plan for reporting unidentified vehicles parked or operated near your building?	❏	❏
Is your parking at least 100 feet away from your building?	❏	❏
Do you limit underground parking to known tenants?	❏	❏
Do you monitor access of trucks to underground loading docks?	❏	❏
Do you have vehicle access deterrents such as barriers, traps, and bumps?	❏	❏
Do you have building barriers, such as cement planters or pillars?	❏	❏
Do you regularly inspect your boundary walls, fences, and gates for any damage or to ensure that security capacity hasn't been breached?	❏	❏
Do you have outside security lighting at your location?	❏	❏
Do you make regular inspections of the equipment to ensure that it is in working order?	❏	❏
Have you considered protecting or eliminating recessed doorways and concealed yards, shrubs, planted areas, and features—anything that could provide a hiding place for an intruder?	❏	❏
Is your facility regularly checked for repairs and kept clear of trash?	❏	❏
Have the locations of outbuildings, bins, shutters, or other potential climbing aids been checked to ensure that they do not offer access to your facility?	❏	❏
Are your trash bins or containers at least 100 feet from your building?	❏	❏
If your trash containers or bins are on wheels, are they locked?	❏	❏
Has your facility been checked at night for outside lighting?	❏	❏
Are all doors sufficiently solid and secured against the possibility of break-ins?	❏	❏
Do all locks, bolts, and other door fixtures meet the necessary security standards for the level of risk?	❏	❏
Do you frequently inspect locks to ensure that they are working?	❏	❏
Are the lock-up procedures for your location handled by competent personnel?	❏	❏
Do you remove all keys to external doors when they are closed?	❏	❏
Are the keys supervised and protected?	❏	❏
Have you thought of using a specialist key handling company?	❏	❏

Table 5–1: Security Survey (continued)

QUESTION	YES	NO
Do you have a way to keep track of who has keys, and to limit the number of personnel with keys?	❏	❏
Have personnel been advised to verify that requests to be on site are legitimate?	❏	❏
Have your windows been checked to ensure that they are secure and fitted with quality locks or limiters?	❏	❏
Do your windows have blast film?	❏	❏
Does your alarm system provide complete building coverage?	❏	❏
Are there alarm systems for your window and emergency egress doors?	❏	❏
Do you inspect and maintain your alarm system regularly?	❏	❏
Are cameras checked regularly to ensure that they are working?	❏	❏
Have you sought professional guidance on the placement of cameras, remote monitoring, and the newest technology?	❏	❏
Are access points to roof level, antennas, wireless equipment, and transmission equipment protected?	❏	❏
Do you monitor and protect outside air entry points?	❏	❏
Do you maximize air filtration?	❏	❏
Are your filters checked and changed regularly?	❏	❏
Do you protect and monitor your water supply, meters, tanks, gas, and power supplies?	❏	❏
Is every employee processed into your building? How?	❏	❏
Is the processing system monitored? Are all keys or cards accounted for regularly?	❏	❏
Are visitors processed into the building using temporary ID cards?	❏	❏
Are visitors or vendors escorted while on the premises?	❏	❏
Is there an identity badge system in place? If so, are there records of all visitors, and are badges only released against signatures?	❏	❏
Do you train your personnel, especially reception, in security awareness?	❏	❏
Does your building screen bags, packages, and briefcases?	❏	❏
Does your building have a Closed Circuit Television (CCTV) system?	❏	❏
Is your CCTV system monitored?	❏	❏
What type of security system is there for the communications center?	❏	❏
Are your computer and radio rooms protected?	❏	❏
Do you have a bomb incident plan?	❏	❏
Have you trained your personnel in the plan?	❏	❏
Do you have an evacuation plan?	❏	❏
Do you have trained search teams?	❏	❏
Are your tower sites protected?	❏	❏

that you can prevent *all* damage, but rather to try and *limit* the damage sustained to the occupants and equipment inside. Film coating is designed to reduce flying glass by keeping sharp glass pieces intact. Polyester film installed on the inside of existing windows performs well under explosive tests. When deciding to install film, keep in mind that the ability to reduce glass hazards depends on a number of factors, including the thickness of the film and how it meets the edge of the window frame. Solar film (2–3 mm) has been used for years as an energy saver, but it is not thick enough to contain glass shards in an explosion.

The thinnest film used for glass fragments is 4 mm. This thickness was used in the 1980s by the United States Army Corps of Engineers and the State Department. Various disadvantages include discoloration with age, surface abrasions, peeling along the edges, and clouding with cleaning. The low initial cost for the material may be offset by the number of times it must be replaced over the lifetime of the building. Newer, thicker film improves on the glass retention properties. Thicker films have been shown to hold more of the material together as the glass exits the frame.

Installation of glass film has also changed over the years. Glass film is traditionally applied to the inside face of the window and cut along the edge of the window with a sharp tool. This is termed as a "daylight" application. There is discussion that cutting along the edge can cause a weakened line where the glass can break at low pressure levels. One way to upgrade the weakened area is to cut the film along the edges of the vision panel and secure it with a bead of silicone sealant. This is considered a "modified daylight" application. There are other methods to fasten the window film to the frame more securely. They are more difficult to do in a retrofit application, as they require the removal of the windowpane from the frame. You can also mechanically fasten the film to the frame using fasteners, battens, or by using anchored stops behind the window. These methods need to be checked to ensure that frame failure does not occur. If the existing frame is not sturdy enough to transmit the explosive load, then you may have to place steel braces behind the frame system or replace the entire window system. There is little reason to have protective film installed if the surrounding anchored frame structure is weaker than the window itself.

A blast curtain concept that may work for your facility would involve the application of window film of 7 mm or greater and a bar or grilling placed behind the window to stop the filmed glass from flying into occupied space. It may not look architecturally pretty, but it can provide windows for your operators and some amount of protection from glass fragments that, in many cases, are the largest cause of injuries during explosions. Glass film can also have a dual purpose of eliminating UV radiation, glare, and heat infiltration. Ballistic glass is also an option, although an expensive one.

Provide alarm systems for windows and emergency exit doors. Windows on the ground floor at the rear of the building are a huge risk for unauthorized entry. All ground floor windows should have substantial window locks. Window bars can be added, but should only be considered where it would not affect an emergency exit in case of fire. Security should not impede escape. Blinds or reflective window film can be used to prevent the easy view of high-cost equipment, communications equipment, or other high-risk areas. Doors should be flush with the building, avoiding recesses. They should fit the frame well enough to prevent forced entry from pry bars. The frame should be as strong and as securely fixed as the door itself. External hinges should be protected and hinge pins made non-removable. External doors and security doors should be fitted with a closer. These should be on the inside face of the door.

Look for neighboring buildings and access points to roof levels, and determine the need to reduce access from those areas. Rooftop assets such as antennas, transmission equipment, wireless equipment, or even emergency generators should be protected from tampering or damage.

UTILITIES AND VENTILATION SYSTEMS

If a biological or chemical terrorist attack were to occur, how long would it take for you to realize your building was affected? By the time the building engineer or person in charge of the HVAC system knows of what has happened, it might be too late to turn off the air handling system to reduce transport of chemical or biological agents through the building. Protection of your air-handling or water systems should be considered in your overall security planning. If there is notice or threat of an attack upon your building, shutting down the system prior to the attack will reduce the movement of the chemical or biological agent throughout the building. Other considerations for the utilities and ventilation systems include the following:

- Outside air entry points should be protected and monitored. Even the introduction of smoke or noxious gases can cause disruption to the communications personnel. Basic security includes cameras and alarms.

- Secure the ventilation system, including all intakes and the network of supply and return duct work, with alarms or motion detection systems.

- Improve the filtration so that it approaches the maximum that the existing system can handle. Ask questions of the system maintenance personnel, such as "Has the filter been changed lately?" An overload in the filter can cause separation of the filter from the screen and allow air to blow past unfiltered. Also ask "If the system has a bag filter, is the bag blown out?"

- Consider installing "scrubbers" that remove biological or chemical hazards from the air duct system.

- Install ultraviolet lights that kill bacteria. Provide pressurized air locks at all entrances. This can keep pollutants such as smoke, fumes, or biological clouds from entering the building before they pass through its intake filters.

- Protect all building water supplies and restrict the access to water meters and tanks and gas and power supplies.

- Put emergency plans that include the building operating engineers into place. Part of the emergency plans can include shutting down the HVAC system and exhaust fans, but this should be in cooperation with the emergency plans and only e done by trained and authorized personnel.

BUILDING ACCESS

Most employees will understand the need for increased security. Discussing the need for building protection with the staff and outlining security considerations will help everyone accept the new and possibly disruptive measures to which they will be subject. Help them to understand it is not just about the building and equipment; communications systems need *people* to operate them and their protection is paramount. There are a number of ways to provide access into the building. Again, as with all of these security systems, you will need to determine the level of security needed and the associated installation and maintenance costs. In a building with few employees, general building access may be controlled by giving each employee a key, a code to an electronic keypad, or a card that operates an electronic access system. The drawback to individual keys is the control and accounting of each key as the entire building or section is at risk if a key is lost, stolen, or misplaced. Larger buildings or facilities with multiple occupants may have other systems to separate employees with valid credentials from visitors, contractors, messengers, and delivery people. In this fashion, people can be quickly processed into the building if they have employee photo ID cards for visual screening or card access control. Visitors can be directed to another line for security screening and inspection. Temporary access cards or time-expiring badges can be issued to preapproved visitors or contractors. Electronic-card access control can also keep track of who is in the building, which is useful during an evacuation or attack when each person needs to be accounted for.

In areas of high risk or concern, metal detectors to screen all persons who enter for guns, knives, or explosives may be appropriate. An X-ray machine can also screen bags, packages, purses, and briefcases. Deliveries by persons or courier services can be directed to a central mailroom or desk for appropriate screening. Closed circuit television systems (CCTV) can provide visual surveillance for building entry areas and isolated parts of the building. A great system with the best technology can fail if no one is there to watch what is happening on the monitor screen. Check with professionals about camera setting, remote monitoring, and the latest technology. Make sure that all cameras are checked regularly to ensure they are in good working order.

BIOMETRICS: THE FUTURE?

Biometrics is defined as a technology that uses behavioral or biological characteristics to measure and determine one person from another.

There are so many things in this world that now require keys, Personal Identification Numbers (PINs), ID cards, or passwords. Getting into buildings, signing on to email, looking for airline prices, and making long-distance calls all require a password. If your security system is good, your building password may change every month. Yet look at many computers and the password into the company intranet is on a small scrap of paper taped to the side of the monitor.

In the near future, all you may have to do is step up and have your eye or face scanned, or put your finger or thumb onto a pad to access your office, computer, or the Internet. If you have an annual pass to Disney World, your pass is inserted into a reader and the first two fingers of your right hand are scanned. This allows Disney to control access to their annual passes and prevents stolen or lost passes from being used. Many companies and organizations are moving toward two or more types of security, such as card readers and fingerprints. Could this technology come to a center near you? The following are biometeric systems that may become the future of security.

Fingerprint Recognition

Fingerprints have up to 50 points of recognition that can be scanned. This is becoming a common way to identify an individual. Problems include worn away prints or cuts and scratches on people's fingers, which would make them unreadable. Fingerprints are mostly associated with criminality, but because eighty percent of 9-1-1 communications centers are located in police agencies, this should not create a problem for employees. Privacy concerns can be alleviated because most fingerprint systems do not store the fingerprint information but simply scan the fingerprints for recognition.

Handprint Recognition

Handprint recognition is far less accurate than fingerprint recognition, but is not associated with criminality. New technology focuses on the inside of the palm, looking at veins and tissues. This is done with an infrared camera scanning the inside of the hand.

Iris Scanning

Iris scanning looks at the front colored areas of the eye. It is far more secure than fingerprints because it has over 260 unique points of reference. It is still a new and expensive technology, and many people are concerned about having their eyes scanned by a bright light.

Retina Scanning

Retina scanning is as accurate as iris scanning, but it scans the back, black area of the eye. This means a much brighter laser is needed and the person must stand closer to the scanner.

Facial Recognition

Facial recognition uses only fifteen to twenty points of reference, making the technology less accurate but also relatively cheaper. There is some use of this technology to scan crowds for known criminals, but a terrorist without a record will not be in the system to be recognized. Facial scanners need good-quality pictures from which to make a positive match. Aging, weight gain or loss, or even differing facial expressions could cause false positives or negatives, creating a situation much like the false alarms from home alarm monitoring systems. Some police agencies do not respond to home alarms anymore due to the high number of false ones. Most facial recognition technologies use a two-dimensional camera, but some companies are working with a three-dimensional camera that measures 200 points of reference in the bone structure of the face.

Thermal Recognition

This technology scans a person's face for heat emissions. This very expensive technology scans 19,000 points of recognition and is not fooled by external temperatures, aging, or even if the subject just finished exercising.

SIGNATURE METRICS

This is simply a system to analyze the shape of a user's signature. It may seem outdated, but it is quite accurate.

VOICE ANALYSIS

This system measures voice patterns. The program must be programmed to your voice, and it takes hours to set up properly. Illness or stress can affect the pattern of your voice, making the system unable to recognize you. Yet mobile communications and telematic systems with voice recognition are gaining in popularity and will become important in the future. Watch for this technology to expand as the communications market expands.

ACCESS TO THE COMMUNICATIONS FLOOR

Consideration must be given to not only the exterior and interior of the building, but also to the actual communications room. Decisions need to be made and implemented to determine who is allowed in and how they will be monitored. Will vendors need a temporary or permanent ID card? Will family or guests of employees be allowed in? Even questions of packages or personal mail being allowed in and opened will have to be addressed. In light of the anthrax threats and concerns, some centers do not allow mail or packages in the communications center dispatch room, thus reducing the risk of a powder or bomb disrupting center operations. Some centers do not allow tours of the communications floor and require vendors to sign in and have their identifications verified before they are given access to the communications room.

This also should extend to the computer and radio rooms of your facility. How are these rooms protected? Do you have a card ID system, key, or code? Some type of system should be in place to create a layer around the actual operations and prevent unauthorized entry. In any case, ask yourself how keys or cards are retrieved from those employees who are fired, retire, or quit (Figure 5-1).

Figure 5–1 Signs, monitored cameras, and keycard entrances will help keep unauthorized people out of your communications center area.

BOMB INCIDENT PLAN AND PHYSICAL SECURITY PLAN

In a large percentage of attacks, conventional explosives are still the preferred weapon of terrorists. Every communications center should have two plans in place to deal with bomb incidents. One should be a physical security plan that covers "hardening" your building to reduce vulnerability. The second is a bomb incident plan that provides a response to a threat or impending explosion to protect property and lives.

Bombs can be made to look like almost anything and can be placed or delivered in a variety of ways. If you are searching the premises for a bomb, look for anything unusual, but let a trained bomb technician decide if it is the real thing or not (Figure 5-2).

Figure 5–2 If you are a communications operator, be one—let the bomb squad be the bomb squad.

Bomb threats are presented as a call-in threat, third party reporting, a written threat, or as a recording. There are a couple of reasons why someone reports a bomb threat. The primary reason is that someone wants to disrupt the daily activities of the communications center. A second reason is that the caller may have first- or second-hand information about the bomb and wants to minimize personal injury or damage to property. The most important issue here is to have a plan and be prepared to react when a threat occurs.

Proper planning and response to an incident will help establish confidence in your management. Having employees understand and participate in the response will reduce panic and let them know that they are not simply sitting in the communications center at personal risk. The decision to evacuate or search is something determined by the committee that establishes the physical security and bomb incident plan. As with your overall building security planning, ask professionals such as police or fire agencies to participate in the planning. If the communications center is the only occupant in your building, then your organization can plan for bomb threats alone. If your communications center is in a multiple-occupant building, each entity should attend the planning meetings. A chain of command for multiple occupant buildings should be established and procedures approved by each agency or occupant. Decide where the command center for a bomb incident will be located. Check with your local bomb disposal unit and determine how to contact them and whether they are available to assist in searching the building during an incident. Representatives from the committee who inspect the building should look for any possible hiding places for bombs and note them on the floor plan. Keep this map of the current floor plan in the command center to utilize during an actual situation.

SECURITY AGAINST BOMB INCIDENTS

Most communications centers have some sort of security in place, such as locks on windows and doors, outside lighting, etc. There is no one plan that covers every threat against a building, but the ability to reduce the vulnerability of your center will make it a safer environment in which to work.

The use of a vehicle to deliver a bomb is a threat not only worldwide, but right here at home. Parking should be restricted to provide a barrier of 100 feet or more from the building shell. If this is not possible, assign spots closest to the building to employees, and have visitors park farther away. Vehicles should be properly marked and identified as belonging to employees.

Planting materials, trees, shrubs, and window boxes or planters make great places to hide bombs. Keep plants low and closely cropped to prevent hiding spaces. The presence of security guards or cameras will also lessen the risk of a bomb being placed.

Post signs that reveal that security systems are being used. Security and maintenance personnel, as well as communications employees, should be on the lookout for suspicious parcels, items, packages, or people. Potential hiding places should be kept locked or inspected on a regular basis. Access doors to sensitive areas or non-public areas should always remain locked with some sort of access system. If any key is lost, the lock should be changed and keys reissued.

RESPONDING TO A BOMB THREAT

Every employee in a facility should be instructed in the bomb threat call protocol, because they can be called into any phone, not just the main emergency line. A bomb threat caller is the best source of information about the threat. The following are guidelines for employees who receive a threat.

- Keep the caller on the line as long as possible
- Ask the caller to repeat the message, and write down as much as you can about what is being said
- If they do not volunteer the information in their initial threat, ask the caller when the bomb is going to go off
- Pay attention to background noises, which might be clues to the caller's location
- Listen carefully to the caller's voice. Is it male or female? Excited or calm? Does the caller have accents or speech impediments?
- Report the information to the proper authorities through the proper channels
- Remain available for interviews by law enforcement
- If a written threat is received, keep all materials, including the envelope

EVACUATE OR NOT?

The most important decision made by management in the event of a bomb threat is whether to evacuate or not. In many cases the decision is made before a situation occurs, generally during the incident planning. How will communication centers function if telecommunicators are evacuated during a bomb threat? The majority of bomb threats are false, yet in today's climate of fear they cannot be dismissed completely. If employees learn that bomb threats have been ignored, they will feel at risk, but if the management evacuates a building every time a bomb threat is delivered, it could also become a problem as possible callers would know how to empty a building and bring communications to a standstill. Students over the years have called in bomb threats to their schools, emptying classrooms right before final exams. Some schools now institute a search, and evacuate only if something is found. Schools have found that a balance must be achieved, as bomb threats were disrupting class time daily. Your building must decide how it will handle a bomb threat before it occurs. Evacuation plans will need to include the use of your backup facility. How will 9-1-1 and communications continue if you leave the communications center? See Appendix F for the Orange County evacuation procedure for relocating the communications center. (Special thanks to Barry Luke for allowing the format for modification and use by other centers.)

Searching a building and then evacuating if a device or package is found is a desirable approach. It offers a response to the incident but is not as disruptive as total evacuation. Once a device is found, prompt and proper evacuation of the premise can be achieved. Employees will gladly evacuate if they know a device has been found. If evacuation is necessary, the floors containing, above, and below the device should be priorities of evacuation efforts. Training for evacuation can be provided by police, fire, or other agencies with expertise. You can train a group of people to assist with evacuation and even use the same people to provide the initial search for a device. A thorough search includes looking in rest rooms, hallways, storage spaces, false ceilings, and any location where a bomb or device could be concealed. Personnel can be trained to look at the building in a predetermined search pattern, marking off areas or rooms as they are searched. The team should understand that they in no way are to handle, remove, or otherwise deal with the device. Proper location of the device should be noted and reported immediately to the command center.

SEARCH TEAMS

It is recommended that more than one individual search each area or room, no matter how small. Searches can be conducted by supervisory personnel, area occupants, or trained explosive search teams. There are advantages and disadvantages to each method of staffing the search teams.

Using supervisory employees to search is the least disruptive and the quickest. There will be less loss of employee time, but it may cause a morale problem if workers find out that a bomb threat was received and workers were not informed of it. Supervisors also may not be totally familiar with an area, resulting in a less-than-thorough search. Using area employees to search is a quick way to determine the presence of a bomb. They are concerned for their own safety and thus will provide a thorough look at their section or area. They will also have the most familiarity of their space in the building so can tell you if something is amiss or out of place. The quicker and more thorough the search is, the greater the reduction in loss of dispatcher or employee time. One drawback to this method is the increased danger to workers. Another drawback is that this method will require all employees to be trained and participate in several practical training exercises.

A search conducted by a trained bomb team is the best for safety and thoroughness as well as morale, yet will take the longest to accomplish. These teams take a considerable amount of time to train and exercise. The decision of who should conduct searches lies with the management and should be incorporated into the bomb incident plan.

SEARCH TECHNIQUES

There are various methods of searching a room. Check with your local bomb team to determine the correct methods to use. The following are a few basic techniques to consider when conducting a room sweep.

The first thing to do in a room to be searched is for both team members to move to various parts of the room and stand quietly with their eyes closed and listen for a clockwork device. Frequently a clockwork device can be quickly detected without the use of special equipment. Even if no clockwork mechanism is detected, the team is now aware of the background noise of the room itself. Background noise may also include rain, wind, or outside traffic noise. The individual in charge of the room sweep should look around and determine how the room will be divided to conduct the search.

The first sweep will be from the floor to a predetermined height. A wall sweep looking at all items on or near the wall. Then together they should look at all items in the middle of the room up to the selected height. During the first sweep, all air conditioning ducts, baseboard heaters, and built in wall cupboards should be inspected within the selected height.

The person in charge then selects the height for the second sweep, usually from the prior height to the top of the head. The team then sweeps the room from wall to wall and all items within that height range. This sweep usually includes pictures, built-in bookcases, and tall table lamps.

The third sweep is made from the head height to the ceiling. This will usually include high-mounted air-conditioning ducts and hanging light fixtures.

A fourth room sweep will be necessary if there is a false or suspended ceiling. This sweep includes flush or ceiling mounted light fixtures, air-conditioning or ventilation ducts, sound or speaker systems, electrical wiring, and structural frame members.

Use common sense in searching. If a threat is made against an operator or center, start searching at the consoles and communications floor and expand the search from there. That would be logical, but do not assume the bomber is a logical person. Continue searching the entire building for suspicious objects. Again, always form a bomb incident plan and follow the pre-designated instructions for evacuation or searching.

SUSPICIOUS OBJECT LOCATED

It is important for all searchers to understand that their mission is to search for and locate any suspicious objects. Under no circumstance are they to move, jar, or touch a suspicious object or anything attached to it. Bomb removal should be done only by trained experts. When an object is discovered, the first step is to report the location and accurate description to the command center, which in turn will notify the police, fire, and rescue agencies. Block off an area of at least 300 feet, including the floors above and below the object. Evacuate the building and do not permit reentry until the device has been disarmed or removed by a bomb squad.

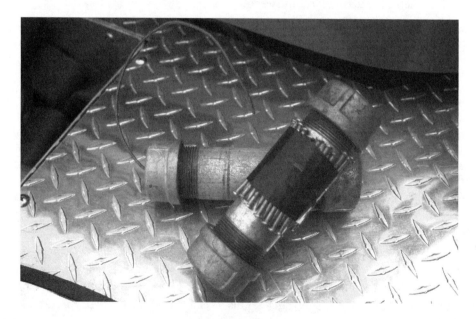

Figure 5–3 The most dangerous explosive device is a pipe bomb, and it is very effective. It is a very dangerous device, not only for responders but for the builder as well. Be alert for suspicious objects.

Remember that this information is only a starting point for looking at your bomb incident plan. Use police, fire, or security experts in developing your plan. Do not leave anything to chance—develop a plan and be prepared.

STEPS IN BOMB INCIDENT PLANNING

1. Designate a chain of command.

2. Establish a command center.

3. Decide what primary and alternate communications will be used.

4. Establish clearly how and by whom a bomb threat will be evaluated.

5. Decide what procedures will be followed when a bomb threat is received or a device is discovered.

6. Determine how communications and dispatching will be handled if evacuation is necessary. How long will it take to activate the backup center?

7. Determine to what extent the bomb squad will be used and when they will respond.

8. Provide an evacuation plan with enough flexibility to avoid suspected danger areas.

9. Designate search teams.

10. Designate areas to be searched.

11. Establish techniques to be used during the searches.

12. Train your search teams.

13. Establish a procedure for reporting that an object has been found and for leading a bomb squad member to the location.

14. Have a plan ready for the chance that a bomb should go off.

15. Establish a simple-to-follow procedure for a person who receives a bomb threat.

16. Review your physical security plan in conjunction with your bomb incident planning.

STEPS IN ESTABLISHING A COMMAND CENTER

1. Designate a primary and secondary command center location.

2. Assign personnel and designate decision-making authority.

3. Establish a method for tracking search teams.

4. Maintain a list of likely target areas.

5. Maintain a blueprint of the building in the command center.

6.   Establish primary and secondary methods of communications. (Caution—the use of two-way radios during searches can cause premature detonation of an electrical blasting cap.)

7.   Formulate a plan for the command center and staff if a threat is received during evening and graveyard shifts, when administration is normally not there.

8.   Maintain a roster of all necessary phone numbers and update it regularly.

EMERGENCY PLANNING AND DRILLS

Every building should not only have emergency plans but also training and drills to ensure that each employee knows what to do if something occurs. Warning systems in multistory buildings should be tested. Emergency messages should be discussed and compared to the emergency plans in place for the building. Practicing emergency evacuation plans can save lives in an emergency. Practice by having announced evacuation drills. These should be planned ahead of time and staff should be assured that it is a drill. Emphasize that if there is no announcement that it is a drill, then people should interpret the alarm as a real event. Sometimes people treat fire alarms like the drills in school each month—as a nuisance. If given adequate assurance that drills will be announced, apathy will turn to proper, fast evacuation over time.

Every plan and security program should be reviewed on a regular basis to ensure that they are operating the way they need to be. If threats or conditions warrant an upgrade to the security of the communications center, then the management or the security team should be prepared to implement increased security levels in a pre-planned method.

NEW BUILDING CONSTRUCTION

If your center is considering new construction or extensive remodeling, then terrorism protection should be an important part of the new building design.

The goal for any new construction is to be protected from all aspects of a terrorist attack and to be the last building standing in any terrorist event. Protection of the employees and the vital infrastructure of emergency communications must be of high priority during building design. Check with local security experts for design consultation.

COMMUNICATIONS TOWER SECURITY

Communications towers are a vulnerable part of any communications system. They are generally located in remote areas away from communities and often are unattended for days or weeks. Security systems will lessen the chance that your tower will become a target for a terrorist attack. A target that is obviously secure is a target that a terrorist is likely to pass by. Generally, criminals will go for the easiest targets: unlocked doors, open windows, etc. The same goes for terrorists. Making your system more defensive will provide more protection from an attack.

A story in the *East Oregonian* in September of 1999 detailed the deliberate damage to the Chemical Stockpile Emergency Preparedness Program. Four ¾-inch cables, which were to activate the system of thirty-five community sirens and nine traffic message boards during a release of toxic chemicals from the Umatilla Chemical Depot, were cut. The intruders also damaged the paging system for the local hospital and sheriff's department. After finding the cut wires, plans were quickly implemented to padlock roads and fences in all tower sites within a week. In the event of a major chemical attack, preventing community notification would add to the panic and possible death toll. This event happened in 1999. Think how the public would have responded if this occurred after September 11, 2001? The public would have put two and two together and predicted a terrorist attack in their own backyard. Imagine the panic today if this happened. So how are your towers protected? In some cases a simple fence will help give a layer of protection from vandalism or attack. There are many ways to add layers to your tower security. Here are a few ideas to consider:

- Razor wire on fencing
- Perimeter fencing
- Twenty-four hour monitored access
- Automatic gate entry system
- Security lighting
- External motion detectors
- CCTV systems
- System alerts that, when tripped, page on-call staff for instant response by staff or law enforcement agencies

Are other systems, such as cellular services, operating on your towers? They could be asked to help provide security systems for the tower. It is in their best interest to secure the site and equipment as much as emergency communications. In any case, discussion of tower security should be initiated. Any discussion should include the "what if" scenario of a tower coming down. How would you operate with one or more towers on the ground? Place the solution into your emergency communications terrorism plan and provide training to your staff. Could a portable tower be placed at the location and function until repairs can be made? Pinellas County Florida Emergency Communications has a portable tower system in case it is needed in such an emergency. Look for federal or state grants to provide this backup for your center if you do not have one already. Security and redundant systems will help your agency continue to operate during a terrorist attack.

SUMMARY

Large buildings or ones with multiple occupants will require a combination of technology and manpower to adequately address security needs. You may need to develop different levels of security for various areas inside of the communications center building. Systems and hardware will not provide total protection, and neither will guards. Both will need to be integrated into a comprehensive security plan. Place those assets at the highest risk within the innermost layer of your security plan. An integrated security system has to be in place *before* an incident occurs, not after. Remember, a false sense of security is the most dangerous environment you can provide.

QUESTIONS

1. Why would communication centers be at risk for a terrorist attack?

2. Who should participate in a security survey?

3. How far away from a building should parking be prohibited?

4. What is biometric technology?

5. How can a ventilation system be utilized by a terrorist during an attack?

6. How should vendors be admitted into the communication center?

7. Name three areas where bombs could be placed.

8. When should a bomb evacuation plan be implemented?

9. Name five ways to protect your communication tower.

10. What two types of plans should your communication center have in place?

CHAPTER 6

Radio Interoperability

OVERVIEW

In this chapter you will see the need to define interoperability and plan for agencies to be able to communicate during a terrorist attack. You will gain a better understanding of public safety and who provides public safety outside of the realm of police, fire, and medical agencies. Communications centers must provide a plan for interoperability that utilizes the Incident Command System (ICS) and provides for training and practice. You will be able to

- Realize the need for interoperability
- Define five categories of public safety services
- Understand that technology is changing rapidly and that the need to provide interoperability increases with the magnitude of terrorist events
- Identify interoperability channels available for planning and use

INTEROPERABILITY: WHEN TERRORISM STRIKES, ARE YOU PREPARED TO TALK?

THE STRUCTURE OF PUBLIC SAFETY IN THE UNITED STATES

The attacks of September 11, 2001 forever changed the way public safety looks at its role and the way public safety responds to terrorist events. Since that date, this country's public safety community has been reviewing its ability to respond in an effective manner. Homeland security is the new term of this century. The public safety community has an unprecedented awareness of the need to communicate in a crisis. Security issues have touched every life in the United States. In this new wave of introspection, agencies are standing back and questioning virtually every aspect of their operations and reviewing their current status and ability to perform their mission, especially to the new threat of terrorist actions. Unlike many other countries, the United States has had the luxury of not regarding terrorism as a primary threat. This has changed.

Public safety is very different in the United States when compared to other countries. Other countries usually support classic public safety from the top down. The federal level pretty well supports all requirements right down to the local level. They provide for the services and capabilities of each discipline that is part of the public safety family. How well they meet the needs of public safety in any given country is variable, but often a single support structure, and resulting processes and procedures are the end product. This greatly simplifies the job of the local public safety service provider, even though it may not serve all of the requirements or even respond well to the needs at the local level.

The United States, on the other hand, has a much more diverse approach to providing the functions of public safety. Yes, there is a federal level with programs and agencies supporting public safety operations, but these are typically limited to actions supporting narrowly defined federal areas of responsibility. The United States Constitution provides for the protection of individual and state rights. The provision of primary public safety services and protection to the average citizen more often emanates from within each state right on down to the local level, not the federal level. The sheriff in many areas is the ultimate responsible entity for providing for the well being and protection of the public. There are 3,066 independent counties in the United States. Cities that provide independent service are also numerous.

While this concept is basic to the United States form of government and way of life, it creates a very difficult environment for any public safety response that spills over the local county or city boundary. Even the responsibility level of similar entities varies greatly by state and from area to area within each state. The position of sheriff in one state may be the ultimate law enforcement and even fire services authority in one area, but in other states, the office of the sheriff may be restricted to serving warrants and other more administrative tasks. State levels of public safety service responsibility vary as well. The state police in one area may have broadly defined authority, while in an adjoining state, the state police are limited to strictly traffic enforcement on state-maintained roadways. The sheer number and variety of responsible public safety agencies in the United States make the term "interoperability" very complex.

The result of such a diverse public safety community makes for an interesting mix of solutions when they are thrown together in an emergency crisis situation. At one time in our past, it may have been sufficient for a single jurisdiction and agency to act in an emergency situation solely within its own area of influence, with limited outside assistance. That rarely occurs anymore. As the United States becomes more diverse, and as the population and economy expands, it is clear that emergency response is becoming more interrelated and complex. The sophistication of infrastructure and the services that the public both expects and demands are putting a severe strain on the public safety institution.

September 11th, 2001 emphasized the need for integrating communications across traditional lines. Clear, efficient communications across traditional jurisdictional and disciplinary boundaries is no longer a luxury, but a fundamental requirement. How an agency responds and prepares for incidents such as terrorism may very well mean the difference between life and death for both the public and public safety workers. In order to respond effectively, agencies must plan ahead. Before they can effectively plan ahead, they must thoroughly understand the nature of the incidents that drive the requirement.

Most agencies are well equipped to handle the requirements of routine day-to-day operations. The communications structure of an agency should reflect those requirements. It is when an incident occurs outside the typical scope of operation that the tears, rifts, and cracks begin to appear. Just because the communications system is sufficient for routine operations does not mean it will perform well under stress. This applies not only to the technical infrastructure, but to operational procedures as well. Most workers work best in a familiar environment. It is the unexpected that tends to throw them off.

The military prepares soldiers to respond almost automatically to a personal threat with very specific responses that have, over time, proven effective in producing a predictable and controllable outcome. This is basic to successfully delivering a response in a crisis situation.

So, what is a crisis situation? How do you plan for it? How do you prepare your personnel to respond to the unknown? Virtually any occasion that is not routine and threatens personnel, the public, or property qualifies as a crisis incident. Obviously, if a crisis incident is of a known and predictable nature, it would be part of the normal routine response and would be "controllable" through the normal process of response.

Law enforcement, fire, and some other public safety operations work in quasi-crisis mode. They never know exactly when a situation will get out of hand and are, to a point, prepared to handle the unknowns. It is when they go beyond their normal support mechanism that a real crisis situation emerges. This is when a standard set of preplanned options is required—this is when working outside the routine needs to have set of routine responses. This is when support from non-routine sources is needed and more importantly, available and predictable—in short, when interoperability is needed (Figure 6-1).

Figure 6–1 Can all agencies in your area talk together if needed?

Recent acts of terrorism have emphasized the requirement for multi-agency, multi-jurisdictional response. Acts of terrorism are not significantly different from any other crisis situation but have one clear distinction. The very act of another human intentionally attempting to inflict damage at random is disconcerting. We have, to a degree, been able to accept nature's random acts and those acts of humans where greed, lust, etc. are the basis for action. We quantify them even when we may not be able to understand them. Random acts of terrorism, by their very intent, are designed to create a panic that goes beyond the act itself. It is the unknown quantity that separates it from other crisis situations. One certainty for crisis situations related to terrorism is that they will certainly require multi-agency, multi-discipline responses by public safety responders. This is why there is such a cry for interoperability in the press of late.

INTEROPERABILITY

Before we can understand the requirements for delivering fully functional interoperability, we should understand and define the terms. There are several critical components that need to be quantified. Demands will be placed on technical requirements, operational requirements, and procedural requirements and they all must be integrated to successfully provide interoperability. To start, we should have a consistent definition of these terms.

Interoperability

Interoperability is an essential communication link within public safety and public service wireless communications systems that permits units from two or more different agencies to interact with one another and to exchange information according to a prescribed method in order to achieve predictable results.

This definition of interoperability was detailed in a document published on a previous September 11th (1996)—the Final Report of the Public Safety Wireless Advisory Committee (PSWAC). PSWAC was a jointly sponsored effort by the National Telecommunications and Information Administration (NTIA) and the Federal Communications Commission (FCC) to help those federal agencies define and document critical needs for public safety communications. The effort comprised a broad range of public safety representatives from across the United States, including agencies of many sizes and disciplines, as well as the vendor community. In Volume 1, the Interoperability Subcommittee (ISC) provided important definitions for the terms "Public Safety," "Public Service," "Interoperability," and "Mission Critical."

DEFINING PUBLIC SAFETY

Even the definition of public safety services is difficult to pin down given the diversity of agencies and localities.

Public Safety

Public safety is the public's right, exercised through federal, state or local government as prescribed by law, to protect and preserve life, property, and natural resources and to serve the public welfare.

The ISC further defined the following terms as subsequent subsets of public safety:

Public Safety Services

Public safety services are those services rendered by or through federal, state, or local government entities in support of public safety duties.

Public safety services are governmental and public entities—or those non-governmental, private organizations which are properly authorized by the appropriate governmental authority—whose primary mission is providing public safety services.

Public Safety Support Provider

Public safety support providers are governmental and public entities, or those non-governmental private organizations that provide essential public services that are properly authorized by the appropriate governmental authority, whose mission is to support public safety services. This support may be provided either directly to the public or in support of public safety services providers.

Public Services

Public services are those provided by non-public safety entities that furnish, maintain, and protect the nation's basic infrastructure and are required to promote the public's safety and welfare.

This latter category is coming under greater scrutiny following the events of September 11th, 2001 as it applies to the provision of Critical Infrastructure (CI) such as wired and wireless telephone service and other traditional utilities such as power, water, etc.

There are those that would limit public safety definitions to the classic description of "guns and hoses," or in other terms, law enforcement, fire services, EMS, etc. While these entities may be the most visible in our day-to-day lives and glorified on TV, others lurk behind the scenes and have just as great an impact both on a day-to-day basis as well as in a crisis. If the police, fire, and rescue personnel cannot get through the streets, then perhaps the street department may be just as important as they are. A similar thing could be said for many other pubic safety and public service providers. The point is, you must clearly identify who the players are before you need to call them. Remember, a local parks service may have emergency medical services personnel on duty every day that may be critical to a successful response.

One thing you can be assured of is that the communications system will be the lifeblood of any large-scale, multidisciplinary interoperability event. It will be the lifeline that links the command structure, the information structure, and the operations structure. It will not work unless it has a preplanned structure. Before you start to create the plan that will work for you, you must reach out and gather some basic information.

GETTING READY

Any successful project has critical stages. Preplanning for major incidents of an unknown type is no different. The first stage is conceptual. It should include a definition of goals, the identification of operational resources, technical resources, and a generalized concept of how those identified resources (operational and technical) could be used to attain the defined goal. Do not rely on a single solution—potential alternatives and backups should be included. Identify which components of the above are mission critical. Identify areas that could cause a catastrophic failure of the mission.

The next stage is creating the plan. Now is the time to get specific. List the potential agencies and services that could conceivably be involved. What radio systems do they have or will they bring with them? Will your communications center need to provide radios for them? Create a slot that includes "others," because there will be potential partners that you may fail to identify on the front end. If they show up, how would you integrate them into your response? Identify categories of responders. Create a template that uses each category of responder to the best of their potential. Let them enter the incident in a role that they are most familiar and comfortable with in their routine day-to-day operations. Do not expect firefighters to be good police officers and vice versa.

The plan should include and revolve around the structure of a standard ICS. There are two very distinct phases in a crisis incident. The very first phase will most assuredly evolve in an unexpected manner and involve any one of your identified responders. Quick action using pre-identified resources (particularly communications) will be critical to containing and preventing damage. The first responder on the scene will fill all of the ICS requirements by observing the incident, analyzing the incident, deploying resources (probably just himself to start), and judging the initial response. The steps are all ICS functions (see Chapter 3 for more information on the ICS system).

As the incident continues, more outside resources will arrive and the ICS functions will expand. Knowing how ICS works and knowing ICS terminology will facilitate responders entering an incident and will help keep control of the response by applying the resources effectively and efficiently. All too often responders will tend to "self-dispatch." It is important for any emergency communications plan to include a clear mechanism for responders attending and entering the event. A staging plan for entering responders is *paramount*. They must be identified and placed into the structure of the incident under the command and control of the ICS command structure. Some responders, ranging from true professionals to off-the-street volunteers, may appear without being requested. There are many issues that point to refusing their assistance, ranging from poorly directed good will and liability issues. With insufficient control, such responders can often become part of the problem rather than part of the solution. Any plans should clearly define how such cases should be handled (Figure 6-2).

The next to last planning stage involves training and practice. If the participants in the real game are not familiar with the ground rules, they will not operate by the ground rules; they will fall back to what is familiar, and that may not be the best choice in a crisis situation. Make the action desired the result of repetition and familiarity. Tabletop exercises are adequate for reviewing general incident flow and

Figure 6–2 Know how ICS works—first responders do.

perhaps identifying possible resources, but nothing beats meeting in the streets. The tactical responders must clearly understand how they "fit" into the scheme of a multidiscipline operation.

The last stage is the one you never wish to see. It is when the crisis situation occurs and the steps discussed are put into action. Even so, it can be gratifying to see a plan work and know you were part of the solution, not the problem.

TECHNOLOGY

All too often we hear of crisis incidents that failed, and the blame is placed immediately on technology. If the technology underlying operability requirements clearly fails, it is more likely due to a failure on the part of planners and managers to identify weak technology or to plan prior to the incident, or technology that is misused or poorly implemented during the incident.

Regardless of this, technology plays a very key role in an emergency incident. It is the base platform and foundation for the operational aspects of interoperability. Radio system(s) and associated components are playing an increasingly important role in today's public safety operations. While we all crave the old simple days, the truth is we could not support our organizations of today with the technology of yesterday. Whether we find ourselves pushing technology or technology pushing us, it is a matter of fact that our communications systems are becoming more complex, bigger, broader, and difficult to manage.

The military has identified the concept of C^3—otherwise known as Command, Control, and Communications—for a long time. This essentially breaks down the critical components of interoperability. Implementation of an incident under an ICS structure is crucial to maintaining a controlled environment with a clear line of command. Providing the communications in support of the first two "Cs"—command and control—is the equivalent of the lifeblood of a response to an incident.

Managing and preparing a communications system to operate in a crisis incident can be very challenging. Some elements can make the task easy, while others make the task difficult and in some cases downright impossible. Radio systems in general have evolved. What was a simple single channel, area wide system (probably only one agency had a radio system) has since expanded and matured. The frequency bands used in public safety radio systems provide a clear case in point. As technology and demand evolved, regulators, manufacturers, and equipment used up frequencies in various ways. Early two-way radio systems operated exclusively in the lower frequencies. As technology and technical abilities advanced, so did the search for a new clear spectrum. Systems expanded and migrated from low band VHF to high band VHF frequency bands, then UHF, and now into the 800 MHz bands (Figure 6-3).

Using radio systems as the base technical platform of interoperability requires technology that is compatible between the players. These systems provide for the "interconnectivity" of an operation. In situations in which responders arrive with radios operating in more than one of the more than four different bands, there is no direct way of communicating. This is not to say radio interoperability is impossible, just more challenging. As stated before, it is essential to identify these hurdles and provide alternatives. In some areas this takes the form of multiple radios in a vehicle. In

Figure 6–3 Do your technicians keep current with technology? Do you provide training opportunities?

some instances, a cross-patch setup is needed in dispatch consoles. More recently, smart gateways, an extension of the older cross-patches, use computer technology to interconnect systems operating on disparate frequencies. Implementing such inter-connections requires preplanning and training.

Regions that have multiple systems in an unshared environment have a few more challenges in order to implement communications interoperability. As previously discussed, there may be multiple frequency bands. Even for those areas that have a common band, identifying specific channels is critical. The best scenario is clearing specific frequencies for only mutual-aid, interoperability purposes. All potential responders must include those frequencies in their radios. Ensuring that all responders know which channels they have access to and how these channels are identified and used is just as important. There have been cases of failed interoperability in which responders could not talk to each other even though they had the same frequencies in their radios. If they do not know which channel to use, or perhaps a channel is referred to by a different channel name, they may never be able to communicate, even though they have fully compatible technology.

Larger systems may have the luxury of many or most responders operating on a shared system. Trunked radio systems have the ability to create "talk groups" that can be authorized for inclusion of all responders on the system. Such talk groups perform the same function of having common frequencies in conventional systems. Some talk groups may be programmed "hot" or available at all times, facilitating direct inter-agency, interdiscipline communications on a day-to-day basis. Others may be programmed into the radios, but disabled until requested through a proper command request procedure. These are local or regional considerations. Typically, as long as all users are part of the same or integrated system(s), this is probably the smoothest technical implementation of communications supporting interoperability.

Several identified interoperability channels are available for planning and use. The 800 MHz band includes five channels set aside exclusively for interoperability. More recently, the FCC has altered its regulations to include several frequencies in both the 150 MHz and the 450 MHz bands to be used exclusively for interoperability. There may be other channels commonly available in your region that would be applicable as well. It should be noted that trunked systems (usually 800 MHz) that use frequencies dynamically and operate through talk groups can be technically interconnected (talk groups) to conventional discrete channels through the use of console patches and other gateways.

SUMMARY

One message to take from this chapter is that technology can greatly facilitate the implementation of an interoperability plan, but can never overcome the lack of an effective plan. Both technology and operational requirements must be considered and implemented jointly and concurrently. You will need to work with multiple agencies to provide the greatest answer to agency interoperability. Many agencies and individuals can assist you in overcoming the obstacle of interoperability. Contact APCO at 1-888-272-6911 or your local frequency advisor. The need is great to communicate during large terrorist events—be ready to handle all of those agencies that will be there to help you.

QUESTIONS

1. How is public safety in the United States different from that of other countries?

2. Define interoperability.

3. What is PSWAC?

4. Define public safety.

5. What is a public safety service provider?

6. List three stages of a successful project.

7. What is ICS?

8. What band has five channels set aside exclusively for interoperability?

9. What is the best scenario for mutual aid interoperability?

10. What is a cross-patch?

CHAPTER

7

Computer Security

OVERVIEW

This chapter will provide a basic understanding of computer systems and security needs. Communications personnel need to understand how security systems work within their computer systems and why things such as firewall protection and virus protection software are needed. Communications centers will see the security holes created by modems and call-in systems and be able to provide security understanding and protection, regardless of how their system is set up. You will be able to

- Identify who should have access to the computer system
- Recognize how the various connections to a communications system can provide security risk
- Understand what a firewall is and how it works
- Understand the value of virus protection software

The moment a user is allowed access to the computer you create a possible security hole. As long as you have people accessing your system, it is impossible to create a system that is 100 percent secure. It is important to begin by determining the needs of your users and then implementing a security plan that fits those needs. A system with too much security can make it too difficult for your users to use. They will stop using it and much needed statistical data could be lost. A system with too much security implemented can also become a management nightmare requiring a very large computer support staff. Likewise, a system with little or no security implemented can make critical data vulnerable to attack, crippling your system at perhaps the most critical time.

PHYSICAL SECURITY

The computer room and the equipment should be in a locked room to ensure that access to the main server equipment is restricted. In the event of a break-in at the center, every locked door encountered would, at the very least, slow down the break-in process and allow time for help to arrive.

Installation of key card type entry, video monitors, and the like will aid in the physical security of your system. Monitoring of the database that is created by access via a key card or other type of secure access, will also periodically assist in determining who is accessing this secure area and why (Figure 7-1).

Figure 7–1 What kind of damage could someone do if allowed access to your computer room? Keep doors locked.

It would be beneficial if the computer room were placed in an area of the building that is not directly accessible, whenever possible, in an area of the building that is not easily located. Personnel will become familiar with those persons they see regularly approaching the computer room and will question those whom they have not seen previously. Your personnel should be advised to "be nosy." If they see unfamiliar people, they should ask them who they are and why they are in that part of the building.

CABLES AND WIRES

There should be a complete cabling diagram detailing where each cable goes. An open cable would easily allow unauthorized access. This detailed cabling diagram should be restricted to only those with appropriate authorization, those with the highest level of security. Access to a diagram is comparable to access to the cabling itself.

Physical access to cables and wires should be appropriately protected. In most computer networks, one cut to one cable could bring the access to the entire system down.

In addition, traffic through the room, even by authorized personnel, should also be restricted, as it is easy to accidentally hit plugs and cause them to come out over time.

Cables and wires should be placed in such a way that they are not running across the floor. This is not only a safety hazard but also could cause them to be stepped on or tripped over, unplugging them from where they are supposed to be plugged in.

AUTHORIZED PERSONNEL

It is important to set requirements for personnel to access your computer system, either by way of logging on to the system or by direct access to the computer room itself. It is important to assess individual needs when determining what level access personnel should have.

It is best for only one or two people to have the administrator password. This password is required for full access to the computer system and enables the user to create and modify user access. Personnel with access to the administrator password are able to upgrade their user access to an administrator level, allowing them easy access to the most secure areas of the system. The administrator password can be sealed in an envelope and kept with your 24-hour personnel in case the administrator is unavailable. If you find that the seal on the envelope has been broken, the password can be changed and the new password put in a new envelope for the next emergency.

ADMINISTRATOR ACCESS

The administrator of the system needs full access to the system. He or she needs the administrator password to log on to the system in order to set up accounts and complete other functions as required by your center. He or she will also need full physical access to the system in order to do required upgrades, backups, etc.

COMMUNICATIONS CENTER DIRECTOR

The director of the center is unlikely to need the administrator password because he or she does not need to modify the users and other components of the system. The director will most likely require full physical access to the system. He or she will want access to be able to view the system and check on those who are making modifications or repairs to the computer room itself.

COMPUTER CENTER MANAGER

Computer managers will need both the administrator password and full physical access to the computer room. It is their duty to be fully aware of all the functions of the administrator. If the administrator needs to be relieved of their duties, the manager must quickly change the administrator password and possibly even the physical locks on the computer room.

COMPUTER HARDWARE TECHNICIANS

Hardware technicians will most likely only need physical access to the system, as their main function is with the physical aspects of the computer system. They will need to have hands-on access to tape drives, cables, modems, etc., all of which are generally located in the computer room itself.

24-HOUR PERSONNEL ACCESS

Since the center is a twenty-four hour operation, supervisory personnel should have access to the administrator password in case the administrator of the system is unavailable. This can be accomplished using the sealed envelope process described earlier. They should also have physical access to the computer room so that they can give access to hardware vendors or to perform required after-hour tasks that may be required or requested by the administrator.

USER ACCESS

Everyday users should have only that access that is required for them to perform their computer tasks. They should have limited physical access to the system, and then only when authorized by the administrator or supervisory personnel.

VENDOR ACCESS

If you use an outside vendor for either your hardware or software needs, it may be necessary to set up requirements for their security level access as well. Remember, it is your computer system. You may want to require that all those accessing your system from an outside company be fingerprinted. You may want to require the exact names and social security or driver's license numbers of those who will be required to access your system. You will then be able to run background checks on the individuals. Even though the company may be reputable, it is the individual employee that will be accessing your system and you need to be responsible for the protection of your system from disreputable employees. Upstanding companies will not have a problem with this.

Software vendors and hardware vendors will only need temporary access to the system and therefore should be monitored by authorized personnel at all times. The administrator can enter the administrator password for a software vendor, allowing them temporary full access to perform software modifications without needing to know the password.

Hardware vendors can be allowed in the computer room on an as-needed basis and should be closely monitored while they perform their required tasks.

A hardware or software vendor should be expected. Hardware or software vendors do not show up unexpectedly. They are specifically requested to come to your site by authorized personnel to perform a specific task. Computer administrators and technicians must notify you that they have requested a software or hardware vendor to come on-site to perform a task. Regardless of what credentials an individual may produce, do not allow them access to your computer system if you have not received prior notification of their arrival.

LOCAL ACCESS

Many public safety computers may be able to allow for only local access—that is, only allowing users who are physically in the same building as the computer system to log on. These users' computers would be directly attached to the computer, and their activity would not pass through any other networks.

This restricts break-in attempts to those who would also have to break into to your building. An individual would have to negotiate the security discussed in the physical security section of this chapter.

A system allowing only local access is much easier to secure. If it is only possible to access the system from within the building, your only security risk is from those who have been assigned access.

DIAL-UP ACCESS

It may be necessary to allow dial-up access, letting users from outside the building use phone lines to access the system from outside locations. This type of access would allow anyone with a computer and a modem to connect to your system. While anyone may be able to connect to your computer, actual access could then be restricted by user logons and passwords.

Even with access restricted to logons and passwords, however, someone with a computer and a modem could break into your system. In time, or with special software, every possible logon name and password combination could be attempted to gain access to your system. Modems come with security options, and with proper setup the modem can be configured to deactivate after a specified number of attempts at the user logon and password. Modems can also have access passwords so that to gain access via the modem, you also must have a password to complete the connection to the modem. Other security measures could include only allowing modem access from certain calling numbers or only during specified times of the day.

If an unauthorized user gains access to your system via modem, they could, in theory, do as much damage to your operation as if they were inside your building.

Authorized personnel should know the location of the modems and of the phone lines that connect to them so that, should an attack occur, they could quickly power down the modem or disconnect the phone line, thereby disconnecting from outside communication.

DIRECT CONNECTION TO THE INTERNET

If your system is directly connected to the Internet, it is possible for anyone anywhere with access to the Internet to connect to your system. Security can then be accomplished not only with user logons and passwords, but also through firewall and virtual private network solutions. Firewall solutions can be as easy or as complex as your needs require. Prices vary from a few dollars to thousands of dollars (Figure 7-2).

Figure 7–2 Are your wireless internet connections secure?

Hackers have gained access to some of the most secure networks in the nation. Sometimes they do nothing more than change a picture on your website or perhaps add text to your web page, making a personal or political statement—anything to get their point across or simply to make the statement that your system is not secure. Some hackers, after accessing a system, have then called the company they hacked into to offer their security services so that the company can benefit from the attack. Some companies have even been blackmailed into accepting services from a hacker who threatens to gain access to the system again and do damage or steal critical data.

Network traffic can also be severely interrupted or stopped by denial of service attacks. This is when the hacker makes so many requests of a computer server that no other valid requests can get through. In this way, they take the network down and leave it inaccessible to authorized users. Your computer system itself may still be up and operational but access to it has been stopped. When this occurs, it is no different than if the entire system has gone down. A system you cannot access might as well not exist.

It is important to know the location of the physical cable that allows outside Internet access to the system. If this cable (or these cables) is unplugged, it would stop the denial of service attack and your system could quickly recover, allowing your local users to again gain access to the system.

FIREWALLS

Firewalls can be as simple as software on your computer system that denies outside access. Firewalls can also be very large and configurable. Larger firewalls must be administrated by personnel trained in firewall security. Personal firewalls can also be loaded on individual computers that connect to the Internet and, with little configuration, can be quite effective. Depending on the solution you choose, you can customize network access, allowing only access from certain networks by port, etc. For example, all network packet data includes the port through which it will pass. A port is like a numbered door down a hall. The port number inside the data packet will then search for a corresponding port on the receiving end. That door can be "locked" by disallowing data to come through that port through the customization of the firewall solution. This is just one way of customizing a firewall. More advanced firewalls can be very complicated. You should acquire the assistance of personnel familiar with firewalls to secure your network.

VIRTUAL PRIVATE NETWORKS (VPNs)

A VPN is a software package that would run on your remote computer as well as the main computer. After a user has established a connection to the Internet, they would start up software on their remote computer and gain access to the network giving them all the functionality they would have if they were attached to the local network. It creates a secure tunnel through which the data is sent to the system. VPNs can also be complicated to configure and manage. You should acquire the assistance of personnel familiar with VPNs to assist you in securing your network.

INDIRECT CONNECTION TO THE INTERNET

A user attached to your system could also have a modem or a second network card in their computer and connect to the Internet. This is in some ways the most dangerous security hole because you may or may not be aware that they are using the computer to connect to the Internet.

The Internet is made up of computers everywhere connected to each other. The computers connect to each other via routes. As long as a computer is attached to the Internet, it knows the route to other computers. Therefore, as soon as a computer connected to your local network makes a connection to the Internet or any other network, all of the computers on that outside network would theoretically know the way or route to all the computers that are connected to that computer. It then becomes a "gateway" or "bridge" to all the other computers on your network.

If someone were monitoring your connection, he or she could sit and wait until your connection is made and thereby gain access to your network. That person would have found a back door into your system. Someone with the right knowledge and patience could severely disrupt your computer network and functions.

It would be best to not allow these types of connections to be made from inside your network. You might allow Internet connections to be made only from computers that are not otherwise connected to your computer network.

ASSESSING SECURITY RISKS

It is as important to be aware of the possible security holes within your system architecture as it is to set up security protections. If you become aware of an attack and if you know how the attacker is making their connection, you should be able to physically disconnect them.

Your system administrator should be intimately familiar with the users that are normally logged onto your computer and at what times they log on. He or she should also be familiar with the processes or programs that would be running on your computers and at what time these programs would be expected to be running. If an attack is suspected, the system administrator can monitor the system and disconnect any users who are not normally logged on during that time. He or she can also abort any programs that are running that would not be normally running during that time.

Your system administrator should also be intimately familiar with the physical connections that are made to your computer. If he or she is unable to determine where the attack is occurring, the administrator could systematically disconnect physical access until the source of the attack can be determined.

It is possible to set up password requirements. Password requirements can be configured so that passwords must be a certain minimum length or longer and must require at least one non-alphabetic character. Passwords can be set up to expire after a certain amount of time, requiring users to change them. You can set up the system to require that passwords not be reused the very next time or that they not be reused over the next twelve required changes. It is important that users be reminded not to write down their passwords and to not give out their passwords to anyone not authorized.

REDUNDANT SYSTEMS

If your budget allows, it would be advantageous to have a "hot" standby. This system would be an exact duplicate of your "live" system. The best way to ensure that this hot standby is truly hot is to switch over to it on a regular basis. Then the system you just left would become the hot standby. If you never use your standby machine, you will not know if it is in good working order.

Administration of a redundant system can be time consuming. Every change that is made on the live system must be duplicated on the other system as well. Depending on the configuration of the system, many changes can be automated so that once it is changed on the live system, the change is automatically sent over to the stand-by system (Figure 7-3).

Figure 7–3 Sometimes a simple sign will help remind personnel about computer room security.

It is the responsibility of the system administrator to make sure that the standby system is completely updated and available for use in case the live system goes down for any reason. There should be a regular schedule for switching to the other system. It is best if the switch to the standby system occurs every thirty days. The system should also be restarted before the switch because many software and hardware problems do not present themselves except during the rebooting phase. This will help the administrator to ensure that the system is always in good working order. Computer systems go through many internal checks only during the reboot phase, and problems can be found and fixed before they become emergencies.

REDUNDANT DATA

If you are on a tighter budget, it is possible to duplicate your data directly on the same system, allowing you to quickly recover your most-used data very quickly. If the actively used data becomes corrupted, you could quickly copy backed-up data on top of the corrupted files. This would quickly get your most critical systems back up and running.

DATA BACKUP

If you are using tapes to back up your data, you should have daily tapes that can be written over each day. On the first day of the month a tape should be removed from the cycle and stored as a monthly tape. At the end of the year a tape should be removed from the cycle and kept as a yearly tape. It is very important to back up your data on a regular basis. Your data should be backed up to media that can be removed from the machine and even from the building. If your data changes every day, you should back up your data everyday. Depending on your operating system you can also make incremental backups of your data that include only the data that changed since the last backup. Your IT personnel will determine use of incremental backups. It is a good idea to have an off-site storage for your backups; if a catastrophic event occurs, you will not lose access to your data. You could bring in a similar computer and restore it with the data from the backup that was stored off-site.

Disk space has become quite inexpensive. Therefore, your system administrator should be provided with a disk large enough that he or she can set up an area of the system as an on-line backup as well. It is much faster to restore data from disk than from tape.

BACKUP CENTER

If your budget allows, it is advantageous to have a completely separate center that your operation could be moved to in case of a catastrophic event. Whenever possible, this system and center should be activated. Unless backups are regularly used and reviewed, you will not be sure that the backup center will be ready in an emergency.

MANUAL PREPARATION

You should have systems in place to go "manual." All personnel should know what to do if there is a total loss of computer systems. They should know what to do to keep the center functional even without the aid of any computer systems or data.

When the system administrator switches to the standby system, the personnel access to the computer systems will probably be unavailable. This would be a good time to test any manual procedures you have in place. In this way, the manual test can be accomplished and any changes to procedures due to failures can be addressed before there is an emergency.

VIRUSES

A virus is nothing more than a computer program designed to damage your system. The program can be instructed to erase all or only certain types of data on your computer. The program can also be designed to change the names of files on your system. Because your programs are written to access data in a certain format, this can be as destructive as if the data had been erased.

Most of today's viruses are received as e-mail attachments. The subject line or e-mail text tricks the receiver into thinking that the e-mail is not destructive. Many viruses are set up to take the e-mail addresses in your e-mail list and forward the virus to them. The virus spreads, and the new receiver opens the e-mail because they think it is from someone they know. Instruct your personnel to never open an attachment unless they are expecting it.

The best protection from viruses is to not allow personnel to receive outside e-mails. If you want to allow your personnel to receive outside e-mails, you could set up computer access that is separate from your main system. However, this can be restrictive in today's environment. If your system is attached to the Internet, you will save a lot of time and trouble by installing virus protection software. Some of the more mainstream virus software programs have automatic online updates that will be downloaded to your computer. As soon as a new virus is created, your virus protection software can be ineffective if you do not have the latest definitions available for its use. If you have an automatic update process available it should be set up to check daily for new definitions and download them to the system.

INCREASED CALL VOLUME

During a catastrophic event, an increase in call volume could cause unexpected problems with your computer system access. It would be advantageous to configure your computer system on the high end in preparation for increased call volume. You may only use a small percentage of your system on a regular basis, but it is wise to make sure that your system has as much memory, disk space, and network bandwidth as possible within your budget constraints so that your system will be prepared for any catastrophic event.

If an increase in call volume causes slow computer responses, it will make this increase in call volume all the more difficult for your personnel to manage. It may be necessary to go partly manual and only use the computer for those calls that require long-term tracking.

SUMMARY

This chapter was not designed to provide technical information because each computer system is unique and different. It does provide a simple explanation of how computer security systems work and why they are important. It is important to ask your programmers to review the security needs of your computer system, find the weak links, and learn how to "harden" your system against a terrorist attack.

QUESTIONS

1. List two ways to physically secure the computer room.

2. Define local access.

3. Define dial-up access.

4. How does a firewall work?

5. Why is it advantageous to have a "hot" standby?

6. What happens when you go "manual"?

7. How does a computer virus work?

8. How can you protect your system from a virus?

9. What can you do when your system has an increase in call volume during a terrorist event?

10. List three communications center personnel and whether they should have access to the administrator's password.

WEBSITES

Check out these web sites for further information about technology security issues:

www.w3.org/Security/

www.mcafee.com

www.norton.com

techupdate.cnet.com/

www.microsoft.com/security/default.asp

CHAPTER 8

Personnel Needs

OVERVIEW

This chapter will identify critical elements of a plan to provide for the personal needs of communications center employees. Communications center managers will be able to identify stressors for communications personnel during major attacks affecting their communications center or community. Personnel will be able to understand the need for Critical Incident Stress Debriefing (CISD) and learn where to find resources. In this chapter you will be able to:

- Identify common communications personnel stress during major attacks

- Develop a family plan for communications center employees

- Understand the elements of CISD and what signs and symptoms operators may reveal when under stress

- Understand the roles communications center employees should have in a personnel emergency plan

PERSONAL NEEDS OF STAFF DURING A TERRORIST EVENT

The personal needs of personnel are often one of the most overlooked factors in planning for a terrorist event. Why should we be concerned about the personal needs of our staff members? The reason is that they are our most important resource. We can have the most sophisticated equipment available, but without a competent, trained staff we cannot mitigate events that have proven overwhelming to communications. Without the ability to communicate properly, chaos rules and renders most public safety operations useless (Figure 8-1).

Figure 8–1 This equipment cannot work without properly trained and cared for employees.

These highly trained technicians are human beings, subject to the same emotions as any average citizen. In a major terrorist event affecting your locale, the operators are going to immediately think of their own families and property. On September 11th, 2001, shock, horror, and dismay hit everyone across our country, including communications center personnel. Stress on the communications operators was also noted during the Oklahoma City bombing. These operators were on duty for long hours and worked closely with the response teams, without being able to directly participate in the disaster work. As we all have been trained to understand, panic is the greatest obstacle to any safe, efficient operation. This is particularly true in public safety service. Taking care of the needs of the communications staff allows the staff to deal with the added stress while providing the highest level of service to first responders and the public.

As you prepare your plans to prevent communications centers from being overwhelmed in the first few minutes of a Weapons of Mass Destruction (WMD) event or natural disaster, this section of the book will focus on some of the subjects, ideas, and issues that will lessen these concerns. These concerns are just the basics. What else can your agency discuss and provide solutions for?

STRESSORS TO BE CONSIDERED

- Concerns about family and property safety
- Overloaded cell phone systems
- Overloaded landline systems
- Passenger calls from hijacked airliner or other unique situation
- Additional notifications beyond the daily routines
- Media requests for information concerning hazardous materials and contamination
- Triage of incoming calls (routine, serious, or urgent)
- Response into "hot zones"
- Call-in system for employees
- Housing and long-term care for additional personnel
- Provisions for care of families
- Integrate plans with other agencies in the best way
- Job ability and event preparedness
- Citizen and responder protection
- Proper coordination with all Primary Safety Answering Point (PSAPS)
- Realistic versus rhetorical response
- Long-term effects of the incident
- Need to handle everyday events at the same time as the terrorist attack
- Length of time personnel can stay at their assigned duties

All of these stressors, and many more depending on your communications center needs, should be addressed in your Standard Operating Procedures (SOPs). These procedures should be provided to each employee, and employees should be trained to implement each SOP. Each employee will then feel more secure about their families, and jobs, as well as their ability to provide effective communications to system users as concerns about continuity and safety are addressed. A big responsibility of communications center managers is addressing concerns and making sure employees know how the system will work during a large incident such as a terrorist attack.

PERSONAL NEEDS

When overwhelming events occur, there has been a tendency for everyone to overlook what communications operators are going through. How long was it after the attacks on the World Trade Centers before we saw any operators interviewed on national television? It was more than a week before anyone realized that the tele-communicators involved in the response did such a great job, yet suffered stress unlike any other responder. They had to deal with the guilt of sending in responders and not being able to raise them on the radio after the towers fell. Emergency communicators do their jobs behind the scenes, every hour of every day, and are often the heroes behind the responders. How many centers have an emergency plan to take care of employee personal needs? After September 11th, several operators were interviewed and asked about their thoughts and needs of the moment.

One single mother interviewed with became emotional and explained that when she left work that day, she went to pick up her ten-year-old daughter from school, and her daughter was crying. Her daughter was very upset and complained to her mom that most of her friends had been picked up almost immediately. She asked why her mother did not come and get her right away. When the mother was asked what she thought she would do in the case of another major attack, she stated that she had very mixed feelings. She understood that her duty was to stay at her console during a crisis, but her natural instinct would be to get her daughter. This is a perfect example of why plans should be developed to care for families. Neglecting them will have an immediate, negative impact on center operations (see Figure 8-2).

Figure 8–2 After a terrorist attack, make arrangements for personnel to speak with their families—it will help them do their job better.

A new committee or staff position could concentrate on handling all of the issues employees will have before, during, and after a terrorist attack. The following are some examples of potential issues.

- Telephone contacts with family members
- A safe location for family members to congregate
- CISD/Employee Assistance Programs for employees and family members
- Proper training to handle WMD and terrorists events
- Drills to confirm proper procedures to assure competence and bring new ideas to increase efficiency of response
- Stock of food, water, and personal supplies for all employees
- Supplies for family members including diapers, formula, and children's food
- Release programs to allow responders to go home in shifts to care for home, animals, and family members
- Areas for sleeping and resting
- Contracting with local hotels and/or restaurants to provide resources beyond the first 72 hours (baloney sandwiches do not taste very good three days in a row!)
- Backup centers in case the main communications center is damaged or destroyed in the attack
- Procedures to provide employee recall to work under specific conditions without the ability to page or phone them

COMMITTEE SETUP

When preparing a communications personnel plan, you will need to form a committee to assist in developing SOPs and determining the supplies that will be needed. Representatives from all working groups within your center should participate in the committee and report back to their respective employees for information dissemination. Within the committee, basic rules and understandings should be agreed upon before planning begins. These concepts will include an understanding that not all ideas will be workable or within the budget. Not everything can or will be budgeted, but with imagination it may be accomplished another way. Everyone may not be completely happy with the procedures, policies, or supplies gathered, but over time the committee will continually accomplish more.

CRITICAL INCIDENT STRESS DEBRIEFING (CISD)

Emergency communications operators deal with life-threatening situations daily. Information is processed and discarded in a matter of seconds. At times this can add up and reach a point that it becomes detrimental to the operator. It could be the build-up of multiple things, or a single incident, that triggers a reaction. This could affect one operator or many, especially in the event of a major incident such as a suicide bomber, a Sarin gas release, etc. During a major incident, operators take phone calls, handle requests from units in the field, handle radio traffic, and deal with what goes on in the communications center—all at the same time. Operators usually deal with the event in their own ways, but it may become necessary to call in a CISD team. In many cases, an incident is reviewed or critiqued without inviting the operator or staff involved. Your agency can send a letter requesting all reviews of incidents include the operator(s) directly involved with the event. As terrorism comes to the United States and the public—unsure of how to respond—continues to call 9-1-1, operators need to understand how CISD can help. CISD is now accepted in fire services as an important aspect of emergency response. It is important that emergency communications staff reach out and ask to be part of the follow-up of any major call or traumatic event that they are involved in. A prime consideration for management must be keeping operators mentally healthy after a terrorist attack in their city (Figure 8-3).

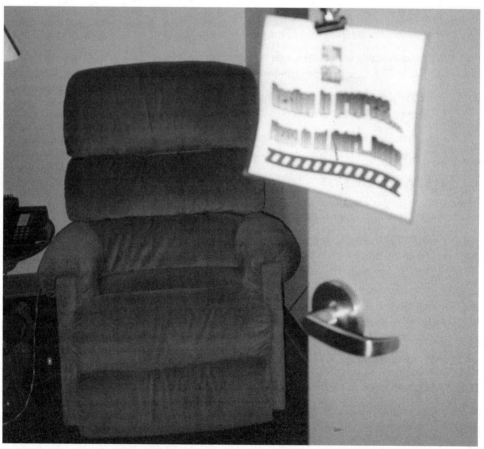

Figure 8–3 Provide an area for personnel to rest and relax during long events.

CISD was created as a tool to prevent or lessen the post-traumatic stress on emergency service personnel. This could and should include the operators. They are at risk of having an adverse psychological reaction to a terrorist incident just like personnel in the field.

STRESS REACTIONS

After an incident is over, operators may experience strong emotional reactions. It is very common, and quite normal, for someone to experience emotional aftershocks after dealing with a horrific event. Sometimes this stress reaction can be seen immediately after or may appear a few hours or days later. There have been cases where it has taken weeks or months before the reaction appears.

Stress reaction signs and symptoms may last days, weeks or even months, and occasionally longer depending on the severity of the traumatic event.

Usually the operator, with the understanding and support of loved ones, will see the stress reactions pass more quickly. On occasion, the traumatic event is so painful it may necessitate the operator to seek professional assistance from a counselor. Operators need to understand that this does not signify weakness, but instead serves as an indication that the event was too powerful for the person to manage.

Table 8-1 and Table 8-2 include some common signs and signals of stress reactions.

Table 8–1: Physical and Cognitive Symptoms of Stress Reactions

PHYSICAL	COGNITIVE
chills	confusion
thirst	nightmares
fatigue	uncertainty
nausea	intrusive images
fainting	blaming someone
twitches	poor problem solving
vomiting	poor attention/decisions
dizziness	poor concentration/memory
weakness	disorientation of time, place or person
chest pain	difficulty identifying objects or people
headaches	heightened or lowered alertness

Table 8–1: Physical and Cognitive Symptoms of Stress Reactions (continued)

PHYSICAL	COGNITIVE
elevated blood pressure	increased or decreased awareness of surroundings
rapid heart rate	
muscle tremors	
shock symptoms	
grinding of teeth	
visual impairment	
sweating profusely	
difficulty breathing	

Table 8–2: Emotional and Behavioral Symptoms of Stress Reactions

EMOTIONAL	BEHAVIORAL
fear	withdrawal
guilt	antisocial acts
grief	inability to sleep
panic	pacing
denial	erratic movements
anxiety	change in social activity
agitation	change in speech patterns
irritability	loss or increase of appetite
depression	hyper-alert to environment
intense anger	increased alcohol consumption
apprehension	change in usual communications
emotional shock	
emotional outbursts	
feeling overwhelmed	
loss of emotional control	
inappropriate emotional response	

WAYS TO RESPOND TO THE STRESS REACTION WITHIN THE FIRST 24 TO 48 HOURS

- Appropriate physical exercise alternated with relaxation should lesson some of the physical reactions.

- Keep busy. This is a normal reaction. You are normal—do not label yourself as crazy. Talk to someone. This can be the most healing medicine.

- Do not use alcohol and drugs to numb the pain; you do not need to complicate this with a substance abuse problem. Reach out. People do care.

- Maintain a normal schedule.

- Spend time with others.

- Help your coworkers as much as possible by sharing feelings and checking how they are doing.

- Give yourself permission to feel rotten and share your feelings with others.

- Keep a journal—write your way through those sleepless hours.

- Do things that feel good to you.

- Realize that those around you are also under stress.

- Do not make any big life changes.

- Make as many daily decisions as possible that give you a feeling of control over your life.

- Get plenty of rest.

- Recurring thoughts, dreams, or flashbacks are normal; do not try to fight them. They will decrease over time and become less painful.

- Eat well-balanced and regular meals, even if you do not feel like it.

WAYS FOR YOUR FAMILY AND FRIENDS TO RESPOND TO YOUR STRESS REACTION

- Listen to them. Spend time with them. Offer help and listening ear if they have not asked for it.

- Reassure them that they are safe. Help them with everyday tasks, household chores, caring for family etc. Give them private time. Do not take their anger or other feelings personally.

- Do not tell them they are "lucky" it was not worse. Those statements do not console people that have been traumatized. Instead, tell them that you are sorry such a thing has happened and you want to understand and assist them.[1]

COMMUNICATIONS RECOVERY

Restoration means regaining a former condition to soften and relieve unlivable situations. After any major event, after-action reports are developed and distributed to the affected agencies. Many large events also produce multiple requests for conference speakers. It is a valuable tool for other agencies to see what happened during a specific event and use the information to review and revise their own policies and procedures. In developing a recovery plan for your communications center, the following concerns must be addressed.

- Revisit all SOPs.
- Replace buildings and equipment.
- Create an alternate site for center.
- Continue CISD.
- Administration should visit with employees to thank and reassure them.
- Schedule for return to normal operations.
- Evaluate need to replace any staff lost in the event.
- Form an alternate plan for long-term temporary operations.
- Budget issues as they relate to overtime and operation costs.

WHY IS IT NECESSARY TO BE PREPARED?

It is the terrorists' goal to undermine the public's confidence in the government's ability to take care of them. We should be prepared to make our employees in communications safe, comfortable, and prepared to handle any large-scale emergency. Knowing what is happening with their family and property and knowing that the administration cares will help operators help their community.

WHAT ROLE SHOULD EMPLOYEES PLAY?

The best emergency plans include employees in the planning process. Specify what staff members should do during a terrorist event and ensure that employees receive proper training on all aspects of the plan. When you include personnel in the planning process, encourage them to make suggestions about center functions and operations during a terrorist situation. Provide all employees with a copy of the emergency action plan. Test not just the action plan, but also the recovery plan.

The plan should address

- Individual roles and responsibilities

---

1. This information was taken from the International Critical Incident Stress Foundation (ICISF) materials and website. For more information on Critical Incident Stress Management, contact the ICISF at *www.icisf.org* or 410-750-9600. They also offer a 24-hour hotline at 410-313-2473.

- Threats, hazards, and protective actions

- Notifications, warnings, and communications procedures

- Means of employee recall to work (besides phones, which may not work) (Figure 8-4)

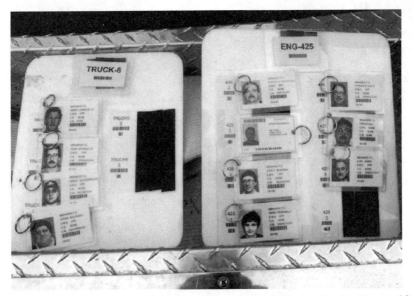

Figure 8–4 How will your center account for your personnel during a relocation or attack? Check with your fire services department.

- Means for locating family members

- Evacuations, shelters, and accountability procedures

- Location and use of emergency equipment

- Emergency shut-down procedures

- First aid procedures

- Prevention of unauthorized access to the center

CAREGIVER, HOW'S YOUR FAMILY?

The following is a reminder to all of us that the role we play in emergency communications also affects our loved ones. This section simply asks you to look at how we treat our families as they become overwhelmed by events we help manage.

Do you remember to find a way to help the people you care about the most? Do you help everyone except those who have done the most to support you? Without the

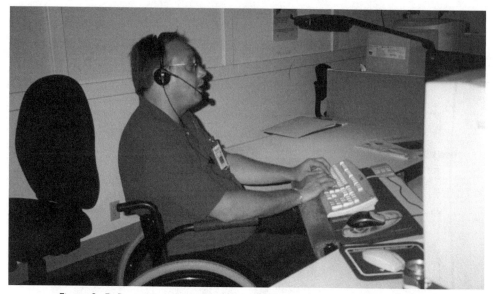

Figure 8–5 Does your evacuation plan include employees with special needs?

help, love and understanding of your family, you would be unable to function at a high level of efficiency.

Think of the many times you missed a special event or cancelled a family gathering to accommodate your work. How many times has your mate had to handle a family crisis because helping others was more important to you?

Remember that your family is always there for you when you stumble or falter in your career. But many times you find it difficult to be there for them when they need you. You want to be there, but you simply are not available.

If you have been in public safety for a while, you have learned to tend to people's physical and psychological needs. This requires you to be upbeat and stay on guard so you will not become too personally involved. When you go home you feel entitled to "down time." You feel you should not have to handle home issues because you have given so much of yourself at work. You resent it when family members want you to help and counsel them during your down time. You are drained, not only physically, but emotionally as well.

Worse, you probably fail to "peoplescape"—that is, to find people who are not in the same field to socialize with. When you are off work, you spend time with work associates. Your conversations and activities take you back to the workplace and you never really get away from it. On one hand you do not want to talk about work when you are at home, and on the other you drag your mate along and talk about work issues you have not shared with them. Of course they feel left out and believe you did not have enough confidence in them to share these things. You do not want to talk about work or what you do. You do not tell your mate many things you feel are confidential or sensitive in nature. You worry that if you tell them something and they accidentally pass it on, it would affect your job.

When your mate has problems (caused by you or other family members) and they withdraw or go into a minor depression, you interpret it as blaming you. Instead of understanding what the real problem is, you become defensive. Your mate is really asking for help to work through whatever is troubling them but you are so caught up in work and your career you fail them on many occasions. All they really needs many times is for you to listen without trying to solve their problems for them.

In most cases you are married to an incredible, understanding mate who takes all of this from you. What can you do now to prevent your significant other from feeling frustrated, left out and unimportant? Your family deserves you at your best when they need you, not just when you are ready.

Many of these things are common to people who work in public safety, especially on shift work. This contributes to a high rate of divorce and attrition in our industry. When domestic problems occur, the affected person can become depressed and unable to function at peak capacity, so it affects not only you personally, but also the workplace.

When you have a strong support group at home, do not ignore it or take it for granted. Remember to take time and make the effort to help your family just as you make time to help others. We guard our innermost feelings from callers and the personnel we serve, and that caution may carry over to our home lives. Try to understand that you probably could not accomplish any of the things you are credited with doing without your family and close friends. Remember that no matter how much you love your job it will never love you back.

SUMMARY

It is important to remember that the greatest asset of any communications center is its employees. They are often overlooked when planning your technology and related issues. We can have the best system in the world, and yet have no one to answer the phones and no one to help the public when they need help the most. We need to plan for employees' continued assistance during a terrorist attack. Knowing their families are safe—getting a few moments to talk to them—may be all it takes to keep operators at their posts. Plan for and hope against a terrorist attack, but be ready to take great care of your employees—they are a critical link in any emergency!

QUESTIONS

1. What is the most overlooked factor in planning for a terrorist attack?

2. What is the greatest deterrent to a safe, efficient operation?

3. List eight stressors to be considered by a communications center during terrorism planning.

4. When will employees feel secure about a terrorist situation?

5. List five items to be developed for an employee emergency plan.

6. Who should be on your personnel planning committee?

7. Why is CISD so important?

8. List ten common signs of a stress reaction.

9. List five ways to respond to a stress reaction.

10. What are some concerns that need to be addressed for communications recovery?

CHAPTER 9

Joint Information Center/Media

OVERVIEW

This chapter will help the communications center's Public Information Officer (PIO) deal with what will happen if a terrorist attack occurs in the center's area and the media starts to arrive. Every communications center should have a trained public information officer who will handle the activities within the Joint Information Center (JIC). The media will be present at all terrorist attacks, and centers should be prepared to provide timely, accurate information to them (Figure 9-1).

In this chapter you will be able to

- Identify what a Joint Information Center does and how it operates
- Know what items to have or bring with you to the JIC
- Understand the Public Affairs Support Function of FEMA as it relates to the operation of the JIC

JOINT INFORMATION CENTERS

The media is a powerful player in any disaster. They can provide the quickest avenue for public information and protection actions. The media can relay safety information, shelter locations, hospital information, road closures, protection-in-place actions, 9-1-1 phone overload situations, and much more. They can, however, provide improper information if a systematic, coordinated approach to public

Figure 9–1 Reporters will find answers to their questions from anyone—a good PIO will be ready to provide them.

information is not provided. This chapter covers what a JIC is and how it functions. It describes the Federal Response Plan actions for the JIC setup. Communications centers should have a trained PIO to assist and function within any JIC set up for a terrorist event. A PIO will provide timely information concerning call volume issues and overloaded cellular problems that we know will occur after a terrorist attack. Training is available through the Emergency Management Institute. Check with FEMA for more information on available courses and dates offered.

A JIC is established to coordinate federal public information activities on scene. It is the central point of contact for media at the scene of the incident. Public information officials from participating state and local agencies also may locate at the JIC. The JIC is a physical location where PIOs responding to an event can co-locate to more closely coordinate information before releasing it to the media in a timely, accurate, and appropriate manner. PIOs may include local government PIOs, local private industry PIOs, state PIOs, and federal PIOs, depending on the nature of the disaster.

There are numerous advantages to setting up and working within a JIC. Some of the benefits include

- Coordinating the release of information from the large number of organizations involved in the response to a terrorist attack.

- Minimizing the amount of conflicting information that reaches the public. It avoids situations in which, for example, the fire department says three people were injured, while the police department says five people were injured.

- Providing PIOs who understand the demands of the media and their schedules.

- Co-locating PIOs to allow for the best use of talents from each person. Some PIOs are better writers, spokespeople, organizers, etc.

- Serving as a "one-stop shopping center" for the media at a site close—but not too close—to the emergency event.

- Conducting new media briefings frequently enough to keep media informed of new developments and provide status reports.

- Providing background data to help the media keep the event in perspective.

- Housing highly qualified, articulate spokespersons to elaborate on and fully explain the terrorist event and local, state, and federal response.

- Making experts available for interviews on topics of particular interest.

- Establishing a well-coordinated and prompt transfer of information from the various command posts to the JIC.

- Coordinating the release of information from the JIC and the other official news routes (part of the joint information system).

At the JIC, steps can be taken to meet the demands of the media. There will be deadlines and live shots to be arranged. In its hunt for information, the press will seek out anything in an effort to get something. Knowing this and working together as a group will allow you to provide accurate and reliable information about what the media needs when they need it. In this sense the JIC minimizes the possibility of conflicting information releases. All information is compiled and confirmed *before* release. No one agency will trump another in getting information out. It will also prevent the press from comparing information and deciding that the information conflicts, thus creating another story of responders not having their act together. Disaster response information should be prepared in advance if possible so that scenarios such as overloaded phone lines, non-emergency calls to 9-1-1, and wireless information is on hand and ready to be given out as the terrorist event unfolds. The JIC is the ideal location for the Rumor Control Center. This is a section that handles all of the truths and rumors that may develop over the time of the terrorist attack. There will be many questions, such as:

- Will there be more attacks?

- What is open or closed?

- Is my family member involved?

- What should I do?

- Can the emergency responders handle the incident?

If the event is a large-scale biological or nuclear attack, it will generate a life of its own and rumor control can be initiated to provide correct answers and update or correct the media if they unknowingly (or knowingly) give information to the public that may incite more fear.

A JIC is not

- The only source the media will use (they will continue to work at command posts and throughout your community to gather information)

- A means of preventing individual departments and command posts from commenting on their own operations

Emergency Broadcast System (EBS) messages should be immediately available to JIC personnel at all locations. This will allow the JIC to provide information to the media and answer any questions about the EBS message.

The location of the center will differ according to the type of emergency. For instance, a site might already be set up for a nuclear power plant attack. A large-scale terrorist bombing may force you to move your designated site to an alternate location. During a terrorist attack that is strictly a local emergency, you might even have to set up operations under a tree until a proper location is found.

Setting up the JIC with the proper equipment and furniture will not pose problems if you plan ahead of time. Again, this can be hampered by the event and situation. If it appears to be a widespread attack within your community, the predetermined location that is ready to be used may be your best bet, even if it is not close to the initial attack.

Information technologies are streamlining the work of the PIO, but remember, any electronic technology may fail when you least expect it and generally at the worst possible time. If your normal route of contact to the media is a cellular phone, rethink how many people will be using their own phone to make calls. During most terrorist events, cellular systems are overwhelmed and unusable as a primary means of communication. Use the latest technologies to your full advantage, but always plan ahead and have alternative strategies and backup procedures. When all else fails, you may have to use runners to relay messages and information between the JIC and the command post.

As a PIO in emergency communications, you need to

- Be ready to cooperatively establish a Joint Information System and a JIC during an emergency or disaster

- Reduce public anxiety by understanding and explaining the local, state and federal response to the media and the public

- Use time in the media to explain when to call for help and to remind the public that 9-1-1 should be used only for true emergencies

Look within the JIC or even within a department vehicle for equipment needed during a terrorist attack to help you efficiently deal with the media. Remember, in today's times, if it even *appears* to be a terrorist attack, you may be inundated with media calls from around the world. Your need for a JIC can be as great during a false alarm as it is for the real thing. The following are some items you may want to acquire for your public information duties.

VIDEO EQUIPMENT

- VCRs and monitors (VCR programmed for local channels)

- Location of VCR, extra VCRs

- External antenna (rabbit ears and UHF loop)

- Supply of blank videotapes (T-160s and T-120s)

- Location of tape supply

- Location of VCR cables

- Tape boxes and labels

- TV program guide

- Dubbing cords

- Instruction booklet

AUDIO EQUIPMENT

- Portable audiotape recorders

- New batteries in the recorders

- Spare batteries

- Extra audiocassette tapes

- Instruction booklet

- Tape labels

- Power cord (if AC adaptable)

- Earphones and patch cords

- Location of spare recorders

FACSIMILE (FAX) MACHINES

- Fax machine

- Group-call capability (to allow a new release to be sent to preprogrammed media fax numbers automatically)

- Answer-back or confirmation feature (gives you a printout indicating the message was received without error)

- Updated list of media fax numbers

- Large supply of fax paper and supplies

- 8 ½ x 11 fax headers/cover pages or adhesive-backed fax identifiers (1 x 3)

- Dedicated fax number

- Fax number directory

- Location of other faxes

- Instruction booklet

CELLULAR PHONES

- Cellular phones

- Adequate supply of fully-charged batteries on hand

- Battery recharger

- Contact information of vendor to supply needed equipment during a large scale event

COMPUTERS

- Computers with sufficient memory to run basic software packages

- Fax/modem capability

- Cellular or standard phone line

- Auxiliary electric power

- Ability to print news releases in the field

- Word-processing software such as WordPerfect or Microsoft Word

- Plenty of formatted blank disks with you and the ability to format blank disks (Always save your working document on a disk after you have saved it to the hard drive.)

Take local courses or use tutorial disks to learn about your computer and feel computer literate. The last thing you need to worry about when you have reporters needing deadline quotes and information is to have your computer equipment lock up or fail. Knowing how your system works will help during those moments of "crisis" when your system has a problem. Also have the following with you:

- Boot disk

- Database program

- Telecommunications/modem software

- Power supply

- Phone numbers and computer service passwords

- Compatible printer, cord, and cables

- Battery and charger (if applicable)

- External modem (if there is no internal one)

- Telephone cords for modem

- Fanfold or other paper supplies

- Ribbons or toner for laser printers

- Location of compatible computers

- All equipment instruction booklets

- Updated release disks

FEDERAL GOVERNMENT ROLE IN JIC OPERATIONS

The following boxed section is from the Federal Response Plan. It outlines the government role during a Presidential disaster declaration. FEMA will participate in the JIC and will provide federal assistance and support. Emergency communications personnel who are the PIOs should understand how the Federal Response Plan works as it relates to the JIC.

PUBLIC AFFAIRS SUPPORT ANNEX

INTRODUCTION/PURPOSE

This annex provides guidance on carrying out the public affairs function in support of the federal government's response to a major disaster or emergency.

SCOPE

The mission of public affairs is to contribute to the well-being of the community following a disaster by disseminating accurate, consistent, timely, and easy-to-understand information. Specific objectives are to

- *Instill confidence that government will conduct response and recovery operations quickly, effectively, and efficiently*

- *Provide critical information about how to apply for assistance and the location and status of life-sustaining shelters and resources*

- *Provide authoritative information to deal with unsubstantiated rumors*

148

POLICIES

FEMA is responsible for implementing federal public affairs activities after a major disaster or emergency. FEMA will develop strategic plans and policies, provide liaison with the directors of public affairs for other Federal agencies and the White House press office, and determine the need for a JIC.

In a major disaster or emergency, a JIC will be established as a central point for coordination of emergency public information, public affairs activities, and media access to information about the latest developments. The JIC is a physical location where PIOs and Public Affairs Officers (PAOs) from involved agencies come together to ensure the coordination and release of accurate and consistent information that is disseminated quickly to the media and the public (Figure 9-2).

Figure 9–2 Your communications center PIO should work closely with on-scene PIOs to deliver accurate information to the public.

A JIC may be established at both FEMA headquarters and/or near the scene of the disaster. Release of information between the two will be well coordinated to the maximum extent possible.

Only one primary federal JIC will be in a major disaster area, preferably co-located with the Disaster Field Office (DFO), either in the same structure or an adjacent structure. This close proximity is designed to facilitate the JIC's access to sources of information about the disaster operation and enable leadership access to the JIC.

Headquarters and on-scene JICs may be established through the execution of other federal emergency operations plans or under special procedures.

Before its release, federal, state, and local disaster information will be coordinated to the maximum extent possible to ensure consistency and accuracy.

All federal agencies may use their own mechanisms for releasing information. No editorial or policy control is exercised by the coordinating PAO over other agencies' release of information about their own policies, procedures, or programs.

State and local governments, as well as voluntary and private responding organizations, are encouraged to participate in and share the resources of the JIC. If co-locating at the JIC is not feasible, all organizations are encouraged to conduct their information activities in cooperation with the JIC.

SITUATION

After a major disaster, normal means of communications in the affected area may be destroyed or severely disrupted; therefore, only limited and incomplete information may be expected from the area until communications can be restored.

The period immediately following a major disaster is critical in setting up the large and complex mechanism that will be needed to respond to the emergency public information and news requirements generated by the disaster.

CONCEPT OF OPERATIONS

Organization

The primary organizational elements of a JIC may vary depending on the size of the disaster and the location of the JIC (headquarters or on scene). Generally, these elements include:

- *The chief spokesperson for FEMA in a headquarters JIC is the FEMA director of media affairs, or a designee, who fields inquiries from national news media. The chief spokesperson in a on-scene JIC is the lead PAO, who may be operating from a Regional Operation Center (ROC) until a JIC is set up in the disaster area. The lead PAO will consult with the FEMA director of media affairs to ensure a smooth transition to field operations.*

- *Media relations, which serves as the primary point of contact for the media for information regarding all disaster response, recovery, and mitigation programs provided by FEMA, the state, and other federal, state, local, and voluntary agencies. This includes providing the media with accurate and timely information on disaster operations to*

> encourage accurate and constructive news coverage, monitoring media coverage to ensure that critical messages are being reported, and identifying potential issues or problems that could have an impact on public confidence in the response and recovery effort.
>
> • Creative services, which gathers information about response, recovery, and mitigation operations and develops and produces information for dissemination by the JIC to the print and broadcast media.
>
> • Multilingual operations, which ensures that non-English-speaking populations receive accurate and timely information about disaster response, recovery, and mitigation programs through appropriate media and in their languages to the extent possible.
>
> • Special projects, which plans and executes projects such as print and broadcast media public-service campaigns, video documentation, surveys, special productions, and logistical support of public meetings and presentations.

HEADQUARTERS-LEVEL RESPONSE STRUCTURE

The FEMA director of media affairs will

- Serve as the White House liaison for all media activities in major disasters and emergencies

- Coordinate public affairs policy, planning, and operations for disaster response in consultation with other agency public affairs directors

- Represent FEMA in a headquarters interagency group, composed of the senior public affairs representatives from each JIC member agency, which comes together periodically to help guide the policies of the JIC and coordinate significant JIC activities

- Manage overall headquarters JIC operations and activities

Federal information released after a disaster is disseminated from headquarters offices through the establishment of a JIC in Washington, D.C., in consultation with the ROC media affairs team, until an on-scene JIC is operational in the disaster area. Depending on space requirements and availability, the headquarters JIC will be located at FEMA, with backup locations available at nearby agencies.

Once the on-scene JIC is operational, it will assume primary responsibility for originating and coordinating federal information released to the media and the public. The headquarters JIC will continue to operate as long as necessary as a satellite of the on-scene JIC and will provide information services to media in the Washington, D.C. area.

FEMA headquarters will provide centralized services to support the JIC in the field. These services include production of the *Recovery Times* newsletter and daily updates for the media, FEMA Radio Network, national media monitoring and analysis, JIC reports, Internet services, and nationwide broadcast fax.

REGIONAL-LEVEL RESPONSE STRUCTURE

The on-scene lead PAO serves as the primary point of contact in the field, handling public information responsibilities in support of the Federal Coordinating Officer (FCO). The lead PAO also oversees hour-to-hour JIC operations (in a particularly large-scale disaster, a JIC coordinator may assume responsibility for JIC operations).

The on-scene JIC should be located where members of the news media are likely to gather. If the DFO is at a remote site, a fully staffed satellite JIC should be established to work with the on-site media. Authority to release public information will remain at the primary JIC (Figure 9-3).

The primary functions of the on-scene JIC are to

- Provide response and recovery information to individuals, families, businesses, and industry directly or indirectly affected by the disaster

- Monitor news coverage to ensure that accurate information is being disseminated

- Take action to correct misunderstandings, misinformation, and incorrect information that appear in the media concerning the disaster response, recovery, and mitigation operations

- Ensure that non-English-speaking populations receive accurate and timely information about disaster response, recovery, and mitigation operations through appropriate news media and, to the extend possible, in their own languages

- Use a broad range of resources to disseminate information to disaster victims and the general public, including the *Recovery Times* newsletter, FEMA Radio Network, FEMA Recovery Radio, Recovery Channel, broadcast fax, and the Internet, as well as traditional print and broadcast news media

- Maintain contact with and gather information from federal, state, local, and voluntary organizations

- Handle appropriate special projects such as news conferences and press operations for disaster area tours by FEMA officials and others

- Provide public affairs support and advice to the FCO and FCO staff

- Coordinate with the logistics section to provide basic facilities, such as communications, office space, and supplies to assist the news media in disseminating information to the public (these facilities are provided as long as the FCO determines their provision to be in the public interest)

Figure 9–3 Satellite broadcast trucks will be on-scene quicker than you think. Will you be ready if they show up at your center?

RESPONSE ACTIONS

Initial Actions

On notification that a major disaster or emergency has occurred, the director of media affairs at FEMA headquarters will

- Contact counterparts at other federal agencies to determine whether there is to be unilateral response to news media or a coordinated response, with one agency serving to articulate the federal response
- Determine the need for a JIC after consulting other federal agencies
- Contact the regional PAO of the affected area (or some other regional official; failing to reach the region, contact the state PAO) to relay information on federal interagency plans
- Coordinate with the FCO and the Emergency Support Team in assigning a lead PAO to deploy to the disaster site and assume public information responsibilities at the JIC

CONTINUING ACTIONS

The director of media affairs will

- Provide advice and support to the Catastrophic Disaster Response Group and keep it apprised of all public affairs actions
- Serve as the focal point for all incoming information from the on-scene lead PAO and JIC
- Ensure that JIC procedures in the *FEMA Emergency Information Field Guide* are put into action

The lead PAO will

- Assume the on-scene lead PAO role on arrival at the disaster area. At that time, the FEMA regional PAO will assume a key management position (either deputy or special assistant, as specified by the Public Affairs Emergency Response Team roster). A JIC coordinator may assume responsibility for the hour-to-hour operations of the JIC.
- Represent the FCO (or FCO's deputy) to the media, public and other agencies
- Serve as an advisor to the FCO and implement public affairs policies and procedures as established by the Director of Media Affairs

Each person representing a JIC member organization will function in two capacities; they will represent the agency in carrying out its public affairs mission and provide public affairs services in support of the various JIC missions.

The following is a real life example of dealing with the media from the PIO of the Tampa Fire Department, Bill Wade.

At times when the national psyche is intently focused on one issue, such as terrorism, even the small impact event can receive exceptional media attention. On September 11th, 2001 the world watched two aircraft fly into New York City skyscrapers. On January 5th, 2002, another aircraft struck a tall American building. In the January incident, however, the following details were different:

- The aircraft was not a jetliner but a four-passenger private plane.
- The pilot was not a foreign national bent on destroying the United States but a "very disturbed" American teenager.
- The building struck was not in a powerful city such as New York, but in Tampa, a mid-size community along Florida's west coast.
- The tall building struck was far less than half the height of the World Trade Center.

There were few real similarities between what happened September 11th and the events of January 5th in Tampa. Media attention from around the world, however, was immediate and quite overwhelming to those involved at the scene and those on the staff of the Tampa Fire and Police Dispatch Center.

For some background, Tampa Fire Rescue (TFR) and the Tampa Police Department serve a city of about 350,000 residents. During the day, due to business and industrial areas as well as a major amusement park, the number of people in the city boundaries can rise to about one million. The fire and EMS combined agency responds to about 53,000 emergency calls every year. The police department responds to about 613,000 requests for service while taking more than 799,000 calls annually.

A modern communications center for the police and fire/EMS dispatch center opened in November of 1999. Prior to this, the two agencies had been housed in separate buildings. In the new building, the fire/EMS dispatch is separated from police by only a partition. The dispatchers share a building but have separate management personnel and operate on different radio frequencies.

On Saturday January 5th, 2002, the TFR dispatch center was staffed with three personnel whose eight-hour shift started at 3 p.m. At about 5 p.m., a fifteen-year-old student pilot took off, without clearance, in a small aircraft from the St. Petersburg/Clearwater International Airport. After coming too close to a passenger airliner, the young man headed towards the airspace over MacDill Air Force Base. MacDill is located on the south end of Tampa and houses the military's Central Command. CentComm, as the military calls it, is where America's war on terror is being coordinated. The young man "buzzed" MacDill and then headed towards downtown Tampa. Somewhere along his flight path a Coast Guard helicopter caught up with him. The guardsmen on the helicopter tried to indicate, with hand gestures, that the young man should land. The young student pilot returned hand gestures indicating that he had no intention of complying. About 5 p.m. that Saturday afternoon, the student pilot's flight and his life ended as the Cessna he was flying crashed into the twenty-eighth floor of a Tampa office tower.

The TFR dispatch center received the first report of the plane crash from the control tower at Tampa International Airport. The tower, in communication with the Coast Guard helicopter, could not provide an exact location of the building that had been struck. Within moments, cell phone callers began deluging the 9-1-1 center with calls. The location of the crash was the forty-three-story Bank of America building on Kennedy Boulevard. The three fire rescue emergency dispatchers knew their shift was about to be very hectic.

With only three fire dispatchers, one person had to work the radio for the units at the incident. The other two people divided their duties between answering non-emergency and emergency phone lines and handling radio traffic for the other alarms within the city. Most of the 9-1-1 calls reporting the aircraft accident were from cellular callers who did not know their exact street location. Of the several dozen callers, most were calm and matter-of-fact in describing what they were seeing.

No one called for help from the impacted building. The building that would have held thousands of people on a business day was essentially empty on the weekend, with only a few dozen people inside.

Initially one alarm was sent to the incident. Upon responding, a fire officer requested an upgrade to a second alarm. The district chief arrived on scene within a few minutes and upgraded the call to three alarms. This sent more than sixty responders to the scene on twenty-one different emergency apparatus. A multiple alarm incident requires contacting off-duty personnel. Using the alpha-pagers for the staff and answering their subsequent calls to dispatch added to the intensity of the workload.

Fire crews arriving on the scene faced multiple problems. The aircraft wings had broken off and crashed to the street, spilling fuel. A large part of the aircraft fuselage was dangling out of the twenty-eighth floor window. There was probably fuel inside the building that could burst into flames any minute, and of course there was concern for anyone who might be injured inside the building or aircraft. On-scene units began setting up staging areas and calling for additional help. Most of the requests were relayed through the fire dispatcher.

Crews on the scene were concerned about terrorism and whether a second aircraft was coming to strike the building. There was also concern that the pilot may have loaded explosives—a secondary device—onto the aircraft. Terrorists use secondary devices to kill or injure emergency responders. After the first event, such as an aircraft crashing into a building, there could be a secondary device hidden somewhere and set to go off when the emergency responders are on scene. A request for bomb squad technicians from both Tampa Police and the FBI was relayed through fire dispatch. No secondary device was found.

Joyce McAlister is a veteran TFR communications technician (dispatcher) with more than fifteen years of experience. She took the initiative to call the control towers at Tampa International Airport and St Petersburg/Clearwater Airport to find out where the plane originated. Initially both towers denied any knowledge of the aircraft, even though the Tampa tower had made the initial report. Eventually the St Petersburg Tower explained the circumstances surrounding the aircraft to McAlister. By this time, on-scene law enforcement had already started piecing the story together. It was nearly 7 p.m. before the fire crews working on the twenty-eighth floor would find out what had happened.

Most of the local news media sent reporters and camera crews directly to the scene. On-scene PIOs from both Tampa Police and TFR coordinated efforts. About an hour into the incident, the national and international news media began paying attention to the events in Tampa. Phone calls from New York, Chicago, and as far away as London and Australia were received at the police and fire dispatch center. If the dispatchers had a minute, they would confirm the basic information for the out-of-town media, but then refer other questions to the on-scene PIOs. The dispatchers relayed requests for information and phone numbers to PIOs through the alpha-paging systems. This kept radio frequencies open for the fire operations. The fire department PIO eventually directed the dispatchers to give out-of-town callers his cellular phone number. This stopped the PIO from having to return long distance toll calls on the cell phone. Because the out-of-town media were given a means by which to get information, it prevented repeated callbacks to the fire dispatch non-emergency lines.

As the incident progressed, each news media outlet had multiple teams of reporters and camera persons working on the story. Each reporter handled a different perspective of the incident. Some journalists worked on the actual ongoing emergency (what fire and police were doing to secure the aircraft, get the aircraft out of the building, assure that everyone in the building and area were safe). Less than an hour into the incident, with the aircraft still dangling from the building, some news outlets asked to talk to those rescuers operating inside the building. PIOs at the scene at the communications center tried to remain current with operations that were ongoing and planned, but usually the information stream was flowing too fast to keep up. As soon as information was confirmed, the news media was advised on events.

The out–of–town media were looking for a telephone interview with a local official. During breaking stories such as this, the media may be satisfied with a recording to get some initial comments from a police or fire dispatcher. The out-of-town caller is typically initially pacified if all the dispatcher can do is confirm the basics, such as what has happened and where. Before an agency allows dispatchers to comment, remember that reporters' questions will not be limited to things the dispatcher knows about. The questions will include: Why did this happen? Has this happened before? Did you have a plan to prevent or respond to this type of emergency? The dispatcher who consents to a telephone interview needs to be briefed on what to say, what not to say, and how to say it. This prevents information from becoming misinformation and facts from becoming hype. Speculation and opinions must be avoided. Only the facts must be stated.

At news desks throughout the Tampa Bay area, assignment editors hung on every word being transmitted to, from, and among on-scene personnel. If a radio transmission was not clear or used jargon that seemed important to the action, there were questions about what was said. It is not the job of the fire dispatch center to keep the media informed of all the scene activity once the PIOs have established themselves on the scene. At the scene, news reporters with handheld scanners would regularly consult with PIOs for clarification and updates.

There were inquiries about the building, its occupants, and the alarm systems. As the on-scene news media talked with eyewitnesses, the stories became vast and varied. One thing is assured at any emergency: if local public safety officials do not talk to the media, reporters will get the story from someone else. As the media talked with the witnesses, some spoke of the fear of terrorism or the lack of safeguards within the building. The news media would then return to the public safety officials on-scene or by phone at the communications center for comment. There would also be inquiries about previous problems with the building. Not having all these facts in front of them, on-site PIOs gave general answers to specific questions and then turned the conversation back to the fact that the scene was stable and work was progressing towards mitigation. At the communications center, dispatchers wisely deferred comment to on-scene officials.

At the fire dispatch center, with everything else going on, media calls and questions were disposed of in a proper and professional manner. If the caller persisted, the dispatcher had the option of placing the caller on indefinite hold or disconnecting the call. During the incident of January 5th, most of the callers were courteous and patient.

As news of the aircraft sticking outside the high-rise spread throughout the Tampa Bay area, well-meaning individuals began calling in to offer assistance. A group of "building-integrity engineers" offered to respond and assess the building stability. A company that owns a very tall construction crane, able to reach the aircraft, offered their aid. One man called and said that the fire department was doing it all wrong and that the plane needed to be brought down outside the building. This caller also stated that if the plane weighed less than 12,000 pounds, he could rig the building and bring it down for us. These offers had to be relayed to the fire officers in charge of the scene and replies from the fire officers had to be relayed back to the companies.

You would think that with titles like communications or PIO, these would be the first places or people to have all the facts about a case. This simply is not the case. This incident involved multiple agencies—local, state, and federal. The incident started in one city and ended twenty-eight stories over the downtown of a different city. It took several days to gather all of the information to answer the questions posed in the first few minutes of the event.

The scene was resolved at about three o'clock the following morning. In the first few hours of the call, fire crews secured the scene by getting people away from the building, placing foam on the aircraft fuel that had spilled into the office space to prevent a fire, and tying off the dangling aircraft using ropes and winches. Plans were made and permission was received from law enforcement and federal authorities to pull the lightweight aircraft into the building. Crews would then remove the pilot's body (he died on impact), cut up the aircraft using basic vehicle-extrication equipment carried by the fire department, and tote the aircraft pieces down the building's service elevator using a hand cart.

By the following Sunday morning there was little left to see. Several broken windows had been boarded up from the inside. A few jagged pieces of glass and exterior façade that were dangling and had yet to be removed hung high above the streets. Below, police kept the curious away so that nothing fell on their heads. Nearby, the out-of-town news media now had an in-town presence as satellite trucks lined a nearby street. For the next few days, journalists from as far away as Atlanta, New York, and Japan searched for every angle of this unique and tragic story. Requests for interviews and records about the incident were directed to the PIOs. To answer the avalanche of questions, the police chief called a press conference Sunday afternoon and had representatives from all investigative agencies present. During the morning and early afternoon Saturday, the fire department PIO worked at the scene with the local, national, and international reporters to answer questions and serve the news media's need to talk with local officials about the event.

Information tends to flow better when the incident is in a single jurisdiction because there is less of it. Sometimes the quick release of confirmed information can stop a small incident from being blown out of proportion. The following is a great example of this.

In the middle of May 2002, the federal government issued a warning that terrorists may be looking to rent an apartment, fill it with explosives, and then detonate the building. Citizens were warned to be familiar with their neighbors and report suspicious activity to the authorities.

On Monday evening, May 20th, 2002 at about 6:40 p.m., an explosion occurred at an apartment complex in an upscale, gated community in Tampa. Within minutes the first fire company confirmed there was no fire, but the windows had been blown out of a second floor apartment. A hazardous materials team from the fire department, bomb squad technicians from the police department, and the local news media quickly headed towards the apartment complex.

The first on-scene crews were told that the apartment was vacant and the prior tenant had been evicted. Police and fire officials treated the scene as if the explosion was intentional, evacuating the building and giving consideration to the possibility of a secondary device.

Several members of the fire department went up into the apartment to look for injured persons and inspect the damage. The origin of the blast was obviously in the kitchen. The refrigerator was heavily damaged, possibly even the source of the explosion. No one was found in the area, and fire crews went outside to report their findings. At this time, maintenance workers from the complex said that they were using insect foggers, sometimes called bug bombs, inside the apartment. They had activated the foggers about six hours earlier. The fire officer on the scene thought that the damage he saw did not match what the maintenance people were saying, and that a couple of insect foggers set off in the apartment would not have caused the refrigerator to explode.

As fire crews worked on-scene, the news media began to address the incident by sending reporters and the infamous news helicopters to the scene. Dispatchers at the Tampa Fire Rescue Communications Center were bombarded with news media asking questions. The dispatchers confirmed the initial information and location but then referred all other questions to the fire PIO, who was already en route to the scene.

The local news media and those living in the apartment complex obviously were wondering about the possibility of terrorism. That speculation grew quickly and, within minutes of the incident, the managers of the apartments were getting phone calls stating that the news was reporting a possible terrorist explosion.

As the news media tried to arrive on the scene, they were blocked from entering the private community by a security guard. The media were kept about three miles away from the site of the explosion. This not only increased their anxiety about what they were not able to see, but it also increased the number of phone calls by reporters trying to find out what was going on.

On-scene fire officers continued to talk with building maintenance about what may have caused the explosion. The apartments were all electrical, so there should not have been a natural gas buildup and no bottles of propane were seen. Finally one of the maintenance people admitted to placing one of the bug bombs inside of the refrigerator. He activated the fogger and closed the refrigerator door. Apparently some meat products had been left to spoil in the refrigerator. The petroleum products that are used as a carrier for the pesticide in the fogger had been allowed to build up inside the refrigerator, creating a flammable and explosive mixture. When an electrical circuit in the refrigerator cycled, the small spark detonated the vapor.

The facts were quickly confirmed on the scene. Then the news was relayed to the fire dispatch center and to the media. Quickly gathering the information, confirming the facts (no matter how unusual), and getting the information out to the news media helps to allay fears and stop some of the hype that can surround small events.

SUMMARY

As you can see, the Federal Response Plan provides a valuable service to your community through participation in the JIC. You should understand how the system functions so that the information relating to the communications system and the messages you would like the public to know can get to the media in the most direct, consistent way.

QUESTIONS

1. What is a JIC?

2. How can the media be used to help the public during a terrorism event?

3. What is a PIO? What is a PIO's primary responsibility?

4. What is an advantage of setting up a JIC?

5. Name two things that a JIC is not.

6. List five things that a PIO should have for the JIC.

7. What is the scope of the public affairs support function?

8. What is a PAO?

9. List four initial actions of the director of media affairs for FEMA upon notification of a major terrorism emergency.

10. How was September 11, 2001 different from January 5th, 2002?

CHAPTER 10

Wrapping it Up for Your Communications Center

THE ROLE OF THE COMMUNICATIONS CENTER DIRECTOR

The communications center director's role in security-related issues is twofold. First, the communications center must be kept as safe and secure as possible at all times. Second, the center must be prepared to handle any emergency that arises—including one that disables the center itself. Prepare your center using the facility security chapter in this book. To accomplish the first task, the director needs to thoroughly evaluate the center's vulnerabilities and make policies to compensate for them where possible.

Directors of new centers often talk about the safety and security measures that were put in place during construction or outfitting of their new centers. This kind of conversation, born of an understandable pride, can create a sort of blindness in which it is easy to believe that because certain risks have been reduced or eliminated, the center is safe from all hazards. This behavior can cause directors to overlook some of the center's vulnerabilities. Anyone who intends to harm your center will not be suffering from this blindness!

One place to start is to look at everything and everyone that comes into your building. This includes the staff and everything they bring in with them; the daily mail; repair technicians; interns, students, applicants and other visitors; supplies and equipment; and even air and water. Anything that enters the facility is a potential point of vulnerability.

Your agency may need to evaluate its policies concerning these factors, or write them if you do not already have them. A number of security policies have emerged at centers nationwide since September 11th in order to deal with some of these factors, including the following:

- Prohibiting employees from bringing unopened personal mail into the center to limit mail borne threats and having center mail opened at a site removed from the center itself to limit exposure.

- Removing outdoor trash receptacles from around the building to remove the possibility that someone could use them as a handy drop-off and detonation place for explosive devices.

- Requiring significant advanced notice in writing from interns and visitors to allow for rudimentary checks and/or notification to appropriate personnel that a visitor is expected (this ties in to the next point).

- Making sure all visitors to the center are known to or expected by the staff before the visit. Do not discount the possibility that uniforms or other insignia from your agency could fall into the wrong hands. Making sure all new hires are introduced to center supervisors and on-duty personnel before they start so they will be recognized can combat this problem. E-mail or paper notification to the staff of the new hire prior to the start date is also a good idea.

- Requiring repair vendors to regularly provide you updated lists of responding employees by name, so their validity can be easily verified, even after hours. Your vendors should also be contractually obligated to provide you with some assurance that keys, etc. to your facilities are safeguarded in the event of employee turnover and that their personnel would pass your state's bonding requirements.

- Prohibiting employees from leaving their ID, uniforms or passkeys etc., in cars or other easily breached places.

Some centers have created policies under which any suspected terrorist or disruptive act in their jurisdiction results in automatic 24-hour security by armed law enforcement personnel who screen anyone coming into the facility and prohibit all deliveries, including the United States mail.

ADDITIONAL FACILITY SAFEGUARDS

It may not be enough to have a great air-filtration system, although certainly that reduces some risks. As a center director, you also need to know the system's vulnerabilities and how to turn the system off and let in outside air if you ever have reason to believe someone has contaminated it. Does your air handler include access to ducts that bypass any filtration system? Similarly, you need to have arrangements in place that allow you to disconnect the water supply to the building and have bottled water brought in, should the need arise.

You should also take a look at your vending machines or coffee service. If someone really wanted to disrupt emergency services in your jurisdiction, what better way than poisoning all your junk food, coffee, or soft drinks? Perhaps it would make more sense from a safety standpoint to have a coffee fund and buy it at the grocery store, or purchase an old vending machine and stock it yourself, using the purchase funds to restock. This measure would protect you from anyone seeking to take advantage of your regular position on a food-stuff delivery list.

Bear these other points in mind when evaluating your facility:

- Do you have an evacuation plan for your facility?

- Does everyone who needs to know about it know about it?

- Do you have a clear idea as to which circumstances merit evacuation?

- Do employees have to call someone for authority to evacuate? If the building's burning, your employees shouldn't have to call you to get permission to leave.

- Know the capabilities of your building—how much wind/weather/water/pressure can it withstand? If the structure is breached, are any hazardous materials released that your employees could be vulnerable to? What kind of backup power generation does it have? What is in the sprinkler system—anything your employees should not be exposed to? Is it earthquake-proof? What force of earthquake? Can it withstand a small bomb? Educate your employees about those capabilities, so they can evaluate evacuating or staying.

- Does your building have any single point of failure? Is there one thing that could go wrong that could take down your CAD, your phones, your power, your backup power, your water supply or your sewer system?

COMMON SENSE IN HANDLING STRANGERS

While no one really wants to create a working atmosphere of suspicion, supervisors and management personnel should probably get in the habit of questioning people they do not recognize who are in the building. Anyone who has a valid reason to be there shouldn't object to this. Many crimes have been committed because people do not question "the guy with the clipboard" because he looks official. Many others occur because people assume that "if he has made it this far into the building, he must be official." People who want to do harm take advantage of these assumptions. A third weak point is that people tend to be naturally non-confrontational. They are therefore eager to accept the first answer they receive when they question someone, as long as it makes any sense at all. Most "bad guys" will have a good answer ready, so it is therefore critical your staff understand that you expect them to follow up the first question with "Who authorized that?" and "Where is your written authorization?" Of course, this means you must get in the habit of requiring and providing written authorizations (except in instances of emergency repair, which your line supervisors will probably be aware of anyway)(Figure 10-1).

As long as we are discussing common sense, let us put in a word about security cameras. Everyone by now either has them or knows they should, but let us face the facts: most people already know they are under surveillance every time they go into a convenience store, let alone anywhere near a government building, so someone who intends to do harm will look incredibly normal and harmless. It is not enough to have security cameras if you do not have the monitors placed at a position that is regularly staffed. Policy should be in place regarding what to do if the person monitoring the cameras sees something suspicious. Guidelines should be created regarding what should be considered suspicious—given that anyone who is up to something will try to look "normal." Your staff's training should include what to look for to trigger a policy response.

GET THE BIGGER PICTURE

These ideas should form part of a comprehensive facility-security plan for your center. Center directors should bear in mind that any attack on their center may be intended either as a distraction or to intensify the effects of another act. The definition of a Weapon of Mass Destruction (WMD) is useful here. According to the FBI, a device crosses the WMD threshold when the consequences of its use overwhelm the capacity of local authorities to respond. Obviously, if a perpetrator can cripple the local authorities' ability to respond in some way, any other device does not need to be as dramatic to overwhelm them.

Therefore, it may be safest to assume for planning purposes that any event that sabotages your center is intended as a distraction from a bigger event. Handling a situation like this might involve deploying your mobile command post or backup center, executing mutual-aid agreements you have with other nearby centers, preparing to reroute 9-1-1 calls (if the original incident does not make that necessary anyway), instituting mandatory call-outs for additional line and administrative personnel, and beginning any special teams notifications, etc. These measures may need to be part of your response plan to any seemingly deliberate incident that significantly compromises your center's security.

You should also assume for planning purposes that your staff members *WILL* have to evacuate. You must know where they should go, how the transition is to occur, and who on your staff requires additional training to make sure the evacuation runs smoothly. Get them that training, and put written evacuation plans in prominent places throughout your center. It is not enough to give these plans to line supervisors to keep in their lockers or desks—what happens if your supervisor is a victim? You

Figure 10–1 Do you really have security if no one is watching?

need to train the people who are most likely to make the decisions (your line supervisors) and then make the information available so that if the decision falls into the hands of your least-experienced trainee, that person will have the necessary resources to handle the situation.

As part of your planning, you probably should invite your special teams' personnel in to do an assessment of your facility so they know in advance where tactical strengths and weaknesses lie. It may be a good idea for all of your special team's personnel to have at least visited your center so they have an idea of what it looks like. You should probably also invite them to conduct training to teach your personnel how to respond if the threat in question comes in the form of hostage-takers.

You may also want to make sure your local FBI office and other potential responding agencies have a copy of your building plan and floor plan. It may be useful for local FBI, etc. to tour your facility. Bear in mind that if an actual event occurs and actually is a distraction for a bigger event, then your own resources, including special teams, are going to be sorely strained. You may have to rely on outside help.

In addition, you or someone on your administrative staff should probably routinely receive electronic copies of daily shift schedules that can be accessed from outside the building. This way you can narrow down how many people are likely to be inside in the event of an incident. You will also have some idea from your personal knowledge of their skills and training what resources you have access to and what you can count on them to know how to do if an incident occurs.

A NOTE

This section is not intended to be comprehensive: it is meant to get you thinking, and to provide some jumping off points. It is not enough to cover what is written here and be done with it—it is important to gather your own personnel and those of agencies you work with in order to make plans and policies that take into account your own center's characteristics. Review various chapters in this book and the security checklist, add your own ideas, and keep looking for ways to improve security.

To that end, once your emergency plans are outlined, look, look, look at them for holes! No better example of this exists than one provided by the events of September 11th, 2001, when one of the phone companies hugely affected by the events tried to implement its contingency plans and realized too late that every single plan involved flying in critical personnel to the affected area. None of them had ever considered the possibility that all air flights might be grounded nationwide, but it happened.

OTHER THINGS TO BEAR IN MIND WHEN PLANNING

As a communications center director, you have to prepare your center on three levels: you have to prepare your staff as individuals, you have to prepare your center as an agency, and you have to prepare your center as a facility. Your plan should be an ongoing, living organism, not a static document that is enacted only when something occurs. Being prepared means everything that can be done in advance is already done when something occurs—and that means starting now, with your training, your facility and threat evaluations, your policies, your procedures and your habits and those of everyone in your agency (Figure 10-2).

Figure 10–2 Denying access to communications center floor involves signage, key cards, and cameras.

The previous section dealt primarily with evaluating and preparing your center, as well as preparing for your center to be unavailable. Below are some ideas to help you get your agency and your staff ready for the unthinkable. These topics are covered elsewhere in this book as well; however, facility preparation is linked closely enough to staff and agency preparation that these bear mentioning here as well.

TRAINING TO AUGMENT YOUR FACILITY'S SECURITY PLAN

Your staff must train in your profession. Training needs to be frequent and ongoing—things change too fast for anyone in our business to get complacent. Even if your budget is extremely restricted, your staff can read industry-specific magazines and books. Other training outlets include seminars and organized courses, conferences and symposia, and agency-specific training available from consultants. Your equipment vendors can probably provide training for your supervisors and technicians as well, so that you have knowledgeable personnel available to deal with equipment issues at critical moments. Moreover, any comprehensive plan for facility security will probably require several different areas of training for your staff, including evacuation training and training in identifying breaches of security.

Remember, disasters will not happen the way you role played them, nor will you get the same disasters for which you planned. You will always get variables, and you have to be flexible enough to accommodate them. Remember the "Theory of Transferability of Training"—the training you do for one disaster will help you with the one that actually happens.

Stage facility-disaster drills, and make them as realistic as possible. For example, get the phone company's permission to overload the phone system, and figure out what you have to do to get around that. (Remember, your center's effectiveness could be breached in just that way—and you should be prepared for it if it is.) See whether your cell calls continue to get forwarded if you set off 100 cell phones at one cell site. Get the affected businesses involved. That way, when something happens, you not only know whom to call to solve problems, and how to reach them, but you also know that person personally, and they understand the mechanics of your problems as well. When the unthinkable occurs, you already have a working relationship with the people you will have to depend on.

Make the drills realistic in other ways, too. Do not have your chief sitting by his phone during a drill, just waiting for your call—he wouldn't be in real life, would he?

Do not buy into "it won't happen here" thinking. Make policies. Get prepared. Then, do not buy the idea that it only happens once. Have a plan A, and a plan B, and a plan C for every contingency. Plan for equipment to fail—even if the vendor swears it will not.

Make sure your backup plans include backup personnel—your chief could be a victim or unreachable. The number two person needs to know everything the chief knows about your response plans and have the authority to execute them. The same is true for the number three person. Are all three ever out of town together? Make sure you have policy to cover that.

OTHER POINTS TO CONSIDER

During a disaster at your facility, there are things to remember and prepare for—the importance of making decisions, for example. As a communications center director, you need to build decision-making authority into your dispatchers and supervisors from the day they are hired, and teach them the kind of judgment skills that foster their decision-making ability. During an incident they may be the only ones you have around to make a decision—and in any case, not every decision can be pushed all the way up the chain during extraordinary circumstances. The kind of decisions you need to consider could range from no longer handling non-emergency calls to delegating tasks to other departments. Yes, mistakes will be made and reevaluations after the fact will occur. But when the unthinkable happens, your people need to act. This requires an act of trust in your employees—make sure you do what you can beforehand to ensure they will merit it.

DEAL WITH THE NEWS MEDIA

We live in a world of instant news now. For example, when the Alfred P. Murrah building was bombed, the news was live within 10 minutes, and live on the scene via helicopter in under 30 minutes. Similar (or even lower) numbers apply to the Pentagon and World Trade Center disasters. Make sure your agency's plans include a media area and a contact (your PIO) and direct the media to it. You may need to restrict the airspace to get control of the media helicopters in some incidents. You can call your local FAA air traffic control tower to get this done. Have the phone number for its supervisor in your plans. You will usually have to articulate why the area of the incident is a "life-safety hazard," and you will probably obtain a restriction to between 2,000 and 5,000 feet. Get it restricted to public safety aircraft only.

PLAN TO HAVE COMMUNICATIONS DIFFICULTIES

What happens if your phones fail, your paging systems fail, and your radios fail all at once? Do your units know to go to an interoperability channel and alert another agency if they cannot reach you at all? Do *all* your units know this, or just the supervisors? Do you have agreements with other agencies that if you cannot be contacted then they should send someone to check on your center, and be prepared to (figuratively speaking) send in the Marines? Does your agency have a plan in place so that if your center cannot be contacted, a coordinated response of your local law enforcement and fire, EMS and hazardous materials (hazmat) teams automatically responds to find out what is wrong? Does your plan include having the other contacted agency dial 9-1-1 into your center and failing to get an answer, does the other agency have the authority to reroute your calls to them as part of a mutual-aid plan? Are you in their plans to do that for them?

Does any response to a signal that something is very wrong include assuming it may not be the equipment that has failed but that some catastrophic hazmat or other event has disabled the personnel?

MAKE SURE YOUR STAFF KNOWS YOUR CONTINGENCY POLICIES...

...particularly who has the authority to call out anyone who is needed.

KNOW WHO HAS THE LEGAL AUTHORITY...

to declare a disaster or request outside assistance (such as the National Guard) if needed. Know who the backup person for that authority is, and the backup for the backup, and keep current contact information for all of them.

In the event that something catastrophic occurs involving perpetrators being present in your center, you may want to have a 10-code or signal in place that your radio operators can use to alert road personnel that there is a problem without alerting the perpetrators. Then have a plan in place for the road personnel to follow, and make sure they are trained on it. (This same 10-code can be used by any unit who gets taken hostage to notify your center the unit is in trouble.)

TECHNICIANS

A big part of any facility security plan involves protecting the communications systems. Most centers, obviously, get their phone and paging services from huge outside vendors, but own or lease their radio equipment locally and have it maintained either by a local vendor or in-house.

To that end, staff charged with protecting technical services and equipment from security breaches probably will need to meet with phone and paging vendors to find out what security is in place, how to tell that it has been breached and whom to notify in that event. It is also useful to have security requirements as part of any bidding process before selecting a vendor to provide these services, if you have enough vendors available to have a choice.

Radio security, however, is likely to be far more involved. When considering the security of your radio system and its vulnerabilities, remember the following points:

- Anyone could thrust an ice pick through a well-chosen cable and do serious damage to your communications capabilities for a given sector.

- Your communications system could be an ancillary victim to an unrelated incident. For example, a plane crash or serious accident could knock over a tower.

- If you lease space to or from other communications entities, or have co-located towers etc., you potentially have much less control over who has access to your equipment, and you must compensate for that contractually and in terms of what you can fence off or otherwise secure at the site itself.

- Any vendor who has access to your equipment should be required to provide you with updated lists of who has access, notify you of any employee who leaves disgruntled and could potentially have made sets of keys, and hire only those who could pass your state's bonding requirements or similar background-check requirements.

- Your radio system's vendor should give you whatever technical information you need in order to form policies to compensate for individual handheld or mobile radios falling into the wrong hands. To form those policies, you should know things like how long portable batteries can be expected to last, whether or not they are readily available (or if readily available substitutes exist) in your area for non-emergency service personnel, whether or not your system allows you to lock out or disable a specific radio, whether or not its location can be tracked (so perhaps it can be found), etc.

- You should have a policy in place for contacting and remaining in contact with your road units in case of a radio system failure, including use of landlines, cell-phones, pagers and dispatching through other agencies via mutual-aid agreements. Your plan should include the possibility that a radio system failure could last twenty-four hours or more. What would you do if your radio system were unavailable for a week—particularly if that became common knowledge, and the local criminal element wanted to take advantage of what it perceived as a period of heaven-sent weakness on the part of the police?

- Assume that some security/operational breaches could occur in the same place as an incident where you were actively working. For example, what if you had large numbers of units deployed to work a forest fire and the tower in that forest that was responsible for most of the local radio reception burned down?

- Do you have adequate redundancy built in, particularly for areas that are likely security targets? For example, if someone wanted to attack a local college football stadium during a game hosting 50,000 spectators and started by taking out the main radio tower that handled that area, do you have adequate redundancy built in that you could still communicate with the site?

- How long has it been since you actually physically inspected your tower sites for security purposes? Are your fences fallen or rusted or full of holes? Are your locks rusted out, or padlocks left unlocked or missing? Do shed doors close properly? Are your sites monitored for security or alarmed? Do the alarms work? Do your own inspections, then ask your repair vendor to rate your sites in terms of security based on his experience. Start with the worst problems and get them fixed.

- If you have a backup radio system—for example, if you have recently switched from VHF or UHF to 800 MHz—are you maintaining your old system in working order? Nothing is worse than to rely confidently on the fact that you have a backup system, only to find when you need it that 1) it has fallen into disrepair, 2) the individual radio units cannot be located or you have insufficient batteries for the radios, or 3) nobody remembers how they work. Keep backup equipment and training current! This includes keeping your repair vendors current on your old equipment. Require in your contract that they be able to take care of both systems. If an "old-timer" at your radio repair shop who primarily took care of your older equipment leaves, ask them what they are doing to train somebody new on the old system. Follow up.

LINE SUPERVISORS

Line supervisors are in some cases the first line of defense. They are likely the most senior personnel after hours. These people are charged with confronting strangers and monitoring security camera information, admitting visitors, and maintaining information regarding evacuations, emergency notifications, and mutual aid. Elsewhere in this chapter many things line supervisors should know have been specifically mentioned. If you are a line supervisor and you do not know them, find out!

The line supervisors' role in center security is threefold. First, as stated, they form a line of defense. Second, it is their responsibility to know the contingency plans and be prepared to put them into action if needed. Third, they must ensure the staff on their shifts knows how to enact plans, and they *must* ensure that others on their shifts know where to find critical information in case they, the line supervisors, are victims or otherwise unavailable. This is not a topic on which supervisors can afford to treat information as turf and protect it. Remember, line supervisors should remember that if they become victims, the information they have shared with their staff may save their lives!

As a line supervisor, get heard. Urge your administration to make plans and get copies of them. Participate in training. Train on a specific topic and offer to conduct in-service to spread the word. Make a point of reviewing the information you are going to have to rely on when the time comes, and note anything that needs to be updated, then follow up.

As a line supervisor, pay attention to who is around your center. Question strangers, or call law enforcement to do so. Check IDs of vendors and visitors. Know who is expected. Report suspicious activity. Pay attention to water that tastes funny, suspicious odors, or strange noises. Many supervisors are somewhat more mobile than line staff—use that mobility to check your surroundings (Figure 10-3).

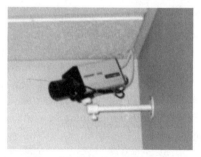

Figure 10–3 Most people are familiar with CCTV systems. Are they monitored by communications center staff?

Pay attention to what *doesn't* happen as well. The phone's not ringing for long periods? Find out why. It is 20 minutes before shift-change and no one has shown up? Is something, or someone, keeping them from approaching the building? You know what is normal for your center. Pay attention to anything that is not.

CENTER LINE STAFF

Center staff may be the first people to know something is wrong—they will be the ones who suddenly do not get the phone calls, do not hear any radio transmissions for a long period of time, or who happen to notice the stranger walking down the hall and notify the supervisors. Your center staff needs three kinds of training: what to look for around the center, how to handle evacuation procedures, and what to look for in incoming incidents that may signal a WMD or other event is transpiring that will affect your center's function and call load significantly.

To address the first type of training, your center staff needs to learn what is abnormal and be instructed to notify line supervisors of strangers, peculiar odors, equipment that functions oddly, and sudden significant changes (in either direction) in call load or radio transmission load. Staff members need to speak up if the water tastes funny, if the vending machine food appears to be tampered with, or if they suddenly have trouble breathing in a particular room or near a particular vent. They particularly need to be alert to oddities concerning visitors. If, for example, they are assigned an intern who takes a seemingly abnormal interest in how secure the center is, they need to alert the supervisor about this (privately, of course).

The second issue is twofold. Line staff members need to know what they should do in case of an evacuation to ensure it runs smoothly, but they also need to know where the evacuation plans are and how they work in case it falls to them to initiate the plan as well. In other words, they need to know their own roles and their supervisors' roles in case of an emergency.

Third, they need to know enough about terrorist acts and WMD to be able to identify potential incidents in the calls they take. Early identification of incidents is critical to minimize the effect both of the incident itself and its ramifications for public-safety communication centers. Many training classes on these topics are now available. We have included the article "Weapons of Mass Destruction: What A Dispatcher Needs to Know." This article contains the basic information to get line staff started.

THREATS

Threats may be more significant from a public safety standpoint than actual incidents because the number of threats is higher than the number of incidents and is growing rapidly. In 1996, there were thirty-six recorded threats of attack. That number has doubled every year since. In 2000, more than 300 threats of anthrax use were recorded, let alone threats to use other agents.

Another reason threats are significant to public safety is they can be just as disruptive as actual incidents. For example, it can take hours or days to determine whether the white powder received in the mail is actually a harmful agent. During that time, those exposed may have to be treated medically as if the exposure were real.

WEAPONS OF MASS DESTRUCTION: WHAT A DISPATCHER NEEDS TO KNOW

By Jennifer Hagstrom, Contributing Editor, APCO Publications

As became amply evident on September 11, the true first responders to incidents involving weapons of mass destruction (WMDs) are often public safety telecommunicators. This piece, based on a presentation given a month before the events of Sept. 11, gives some background on WMD incidents in the United States and some basic guidelines for the public safety agencies who may be required to handle those incidents. What are appropriate responses to these incidents? What do dispatchers need to know about WMDs?

TERRORISM

WMDs are usually the result of some form of terrorism. FBI defines terrorism as "the use of violence and threats to intimidate or coerce for political purposes."

Before September 11, the threat of terrorism here seemed real but not imminent. With the exception of a few major incidents, such as the 1993 World Trade Center bombing and the destruction of the Murrah building in Oklahoma City, terrorism seemed relatively isolated and confined to small, localized areas. Since September 11, however, the spotlight on terrorism has been relentless. Even so, the numbers may surprise you.

How pervasive is the threat of terrorism? How many terrorist incidents have actually occurred?

In 1990, according to the FBI, seven domestic incidents were classified as terrorist acts. In 1993, we had the most in any year during the 1990s, with 12 incidents reported. In 1999, we had 10.

The FBI further reports that in addition to actual numbers, it had, for example, two incidents in 1997 where terrorism was suspected but not confirmed, and an additional 21 incidents where terrorist acts were prevented. Twelve incidents were prevented in 1998, and seven in 1999.

The significant factor in these preventions is that the preventor often was not the FBI or some other federal body—but instead might have been local law enforcement.

An example of a prevented act occurred in Port Angeles, WA in 1999, where an individual was noticed acting suspiciously as he tried to gain entry into the United States. A search of his vehicle revealed a liquid that at the time was believed to be a narcotic. On further investigation, the liquid was revealed to be a highly explosive nitroglycerine-like substance, and the man had additional quantities of it already in a hotel room in Seattle. It is believed that the Space Needle was the intended target, and Seattle authorities cancelled a New Year's Eve celebration there. It is now believed the suspect in this incident may have been working in Osama Bin Laden's terrorist network.

The number of international terrorist acts, obviously, is considerably higher. In 1997, there were 666 attacks. In 1998 the number dropped significantly, but it increased again in 1999.

WMDS

What, exactly, is a weapon of mass destruction? It is any device that releases chemical, biological or life-threatening levels of radiological material. According to the FBI, a device crosses the WMD threshold when the consequences of its use overwhelm the capacity of local authorities to respond. According to this definition, the FBI recorded 12 actual WMD incidents in 1999.

NBCS

Any discussion of WMDs soon brings us to another acronym—NBC. No, it's not the Emmy Award-winning television network. NBC in this instance stands for "nuclear, biological and chemical"—the three kinds of destructive agents commonly used in WMDs.

Some resources list 130 different chemical and biological agents that are available now. Some are extremely deadly, while others are just disruptive.

SOURCES OF NBCS

Agents used in WMDs come from a variety of sources. Considering this nation's mass-transportation system and the information available on the Internet, these agents are increasingly easy to procure or manufacture.

Nuclear material is, obviously, available at nuclear power plants. But it's also commonly present at hospitals and universities and at many corporations as well. Nuclear material is often transported on our interstate highway system or rail system—for example, the contaminated material from the accident that occurred at Three Mile Island in the late 1970s was removed via rail. This raises questions for local public safety agencies regarding their own vulnerability to accident or terrorism, and it also means that virtually no jurisdiction is free of the potential threat.

The collapse of the Soviet Union presents us with several problems. First, their program employed 6,000 scientists who had to seek other employment after their government jobs disappeared, and not all of them can be presumed to be working for the "good guys." Second, the collapse of the Soviet Union's military calls into serious question the security of their stores of biological and chemical agents—who has control of all that smallpox now? Given the shaky economy experienced former Soviet republics since the fall of the Iron Curtain, how can we not wonder what's been for sale to the highest bidder?

Chemical and biological agents are relatively easy to manufacture. Some of these agents are so lethal that one ounce could kill 2.2 million people. The good news is that the delivery methods of these agents have not been successful or are extremely difficult to accomplish. Chemical and biological weapons are so readily available that they are occasionally referred to as the poor man's nuclear bomb. A rule of thumb regarding chemical and biological agents is that as the lethality of an agent increases, the technical knowledge necessary to pull off an attack using that agent usually also increases. Conversely, as the lethality decreases, the knowledge necessary to use an agent decreases.

AVAILABILITY AND PREPAREDNESS

You can get an idea of our potential vulnerability by considering just one biological agent—smallpox. Virtually wiped out as an illness worldwide during the 1970s, small supplies of smallpox have been maintained for research purposes by a few governments. This country stopped routinely vaccinating all children for smallpox by 1980. Those who were vaccinated are believed to need boosters in order to maintain their resistance to the disease.

However, according to a Russian scientist who defected to this country, the former Soviet Union developed the most sophisticated biological and chemical weapons program in world history. They were weaponizing smallpox and anthrax, among others diseases, and at one point were attempting to weaponize the AIDS virus. According to the defector, Russia currently has 800 million pounds of the smallpox virus on hand.

NBC incidents have occurred on U. S. soil. In 1994, a political group in Oregon poisoned 751 people with salmonella by spraying the agent on salad bars in local restaurants in an effort to sicken enough people to sway a local election slated for the next day. The fact that the act was deliberate wasn't discovered until someone involved was later arrested on another charge and confessed.

In 2000, protesters in Minneapolis took diluted cyanide, released one bottle of it in a McDonald's restaurant, and threw others at police officers. Fortunately, the substance was so diluted it did not cause any serious injuries.

This incident received virtually no media coverage—the opportunity for a wake-up call was lost. But Harrison points out that public safety cannot afford to have an "it can't happen to us" mentality, regardless of the size or location of a jurisdiction.

Threats tend to fall into several categories, including the following:

Personal

Personal threats involve someone threatening a specific other person, as in "jilted lover" situations. For example, in November 1999 a woman opened her mail, and some white powder spilled out along with a note that read "you've just been exposed to anthrax." In this instance, the sender was the woman's boyfriend's former lover.

Incident to Another Crime

WMDs are now occasionally being used as the weapon of choice during the commission of another crime—or example, in bank robberies in which the suspect claims to have an NBC weapon.

Terrorist Acts

An example of a threat in this category occurred in Knoxville, Tennessee, in January 2000, when a security guard handling the mail at a Planned Parenthood office opened a powder-laced letter claiming "you've just been exposed to anthrax."

Regardless of type, these incidents are considered federal crimes and the FBI has the primary authority to investigate threats or actual attacks involving these substances or weapons.

THE SUSPECTS

Suspects, like incidents, tend to fall into several categories, including the following:

Lone Individuals

These are typically the most difficult to detect. Lone individuals often lack the funding available to larger, more sophisticated groups, so they often resort to threats in lieu of actual attacks. This is not necessarily always the case, however. Remember the Unabomber?

Local Terrorist and Non-aligned Groups

These groups may have funding. They may also be able to build or purchase an NBC weapon. They can be difficult to detect because they are members of society and fit in. An example of a group like this that is very active is the Earth Liberation Front (ELF), which has been burning up ski resorts and houses in the Pacific Northwest to protest urban sprawl.

Hate Groups and Patriot Groups

These are usually local. In 1999, 435 patriot groups were known to be active in the United States. However, that number dropped to 194 in 2000. The number of patriot groups peaked after the Branch Davidian incident in Waco, Texas, but has been declining since the Oklahoma City bombing. The hate group numbers tell a different story—457 active hate groups were known in the United States in 1999. In 2000, the number reached 602 and continues to escalate.

Edge cities or suburbs can be especially vulnerable to hate crimes if they have had a recent influx of minorities. Also, individuals under the age of twenty-two carry out more than 50 percent of all hate crimes. Often these incidents get the "they're just kids" response, which can be both naïve and dangerous if it blinds authorities (not to mention parents!) to a potential larger, structured network of activities underneath—often until a more serious escalation of activity occurs. Sometimes the group is not a threat, but one lone individual in the group who does not think the group is doing enough becomes a threat. The man who committed the shootings at the Jewish community center in California several years ago is one example (a disgruntled member of the Aryan Nation).

The Internet is bringing hate to people who otherwise would not have found it. For example, a man named Benjamin Smith began to learn about hate on the Internet, and eventually shot at 32 people, wounding eight and killing two, including an African-American college basketball coach. This incident occurred in Skokie, Ill.

There are currently 366 hate sites on the Internet, according to Harrison.

International Groups

One reason why international terrorists can be such a threat is because they often have ample funding and ready access to technologies, facilities, and the necessary technical support to pull off an NBC or WMD attack. Seven countries are considered terrorism sponsors: Libya, North Korea, Syria and Cuba (which are believed to have no direct involvement for 10 years) and Iran, Iraq and

Sudan (who are believed to still shelter terrorists but were also believed to have cut back on their activities). In the wake of Sepember 11th, 2001, these assumptions may have changed, as might have the list of countries.

Doomsday Cults

These cults do not always attract only those people who are bent on self-destruction. For example, the sarin gas attack on the Tokyo subway system, which killed 12 people and injured 5,000, was the brainchild of a doomsday cult leader. This group also tried to pull off an anthrax attack but was unsuccessful.

COMMON NBC AGENTS

The following is a list of some common agents, although it is by no means exhaustive:

- *Sarin gas—Sarin gas is 200 times more lethal than chlorine.*

- *Cyanide—This agent smells like burnt almonds. It was used in the 1993 World Trade Center bombing, but the device failed.*

- *Pepper spray—Obviously, this agent is readily available to almost everyone.*

- *Anthrax—This agent is highly lethal, but the person-to-person transmission effectiveness rate is low.*

- *Plague—This agent has a high rate of person-to-person transmission, and is extremely lethal if untreated.*

- *Smallpox—This agent has a high rate of person-to-person transmission and one third of its victims die.*

- *Ricin—This agent is highly lethal. Ricin is produced from castor beans. In the United States, a prison that allowed its prisoners to grow vegetables was nearly the target of an escape plot using "home-grown" ricin. The plot was foiled only because the prisoner behind it did not keep it a secret; he was turned in by a fellow inmate. (Are your prisons allowing prisoners to grow gardens?) This agent has also been intercepted coming in from Canada.*

CALL-TAKER TECHNIQUES

Harrison offered call-takers the following guidelines and hints, both to help them handle suspected incidents and to give them an idea of what to look for in an attempt to identify an incident:

- *As always, remember the basics. If you suspect something is not right, warn your responders of a possible hazmat incident.*

- *Look for a pattern. If it is a chemical attack, 9-1-1 may get the first calls that something is not right. If it is a biological attack, the hospitals may be the first to know, or possibly EMS responders. Some agents take two or three days for symptoms to manifest, after which hospitals and EMS transport units may suddenly be flooded with people experiencing the same symptoms. On the EMS side, public safety communicators may be the first to establish a pattern. For example, each of nine ambulances may transport one or two patients—not a pattern for them—but all of the related calls may come into the same center, indicating a possible pattern.*

- *Look for a higher than normal number of dead animals and birds—they will usually show symptoms before humans do.*

- *If a suspected device is still present, instruct callers not to touch it. Remind responding units of the potential hazard as well.*

- *Chemical agents usually have very unique signs. One of them shared by many chemical agents is odor, such as newly mown hay, apple blossoms, garlic, geraniums, etc. Look for odors that are out of place. One of the first clues in the incident that occurred in Minneapolis was the odor of hay in a McDonald's—very out of place. Call-takers can ask about out-of-the-ordinary odors, but they should remember that the absence of an odor does not rule out the possibility of an NBC agent.*

- *Stage incoming units until the hazmat team arrives. For some NBC agents, the typical gear (including a self-contained breathing apparatus) that firefighters wear is sufficient, but not for all. However, the responders cannot know whether what they are wearing is sufficient until it is already too late to do anything about it. The safe option—waiting for the HAZMAT team—is the best.*

SEPTEMBER 11TH, 2001: A WAKE-UP CALL

At the time of Harrison's presentation, no one could have predicted how timely a topic terrorism would become in the United States. One of the most chilling aspects of the events of September 11th is that the weapons used—airliners—were not typical WMD or NBC devices. It did not require any detective work to determine that the events were terrorist acts. Still, public safety communications professionals have been given a tragic and very clear wake-up call. Now more than ever, they have a responsibility to be prepared.

Based on a presentation by Tony Harrison, Public Safety Group, given at the 67th Annual Conference & Exposition in Salt Lake City, Utah, August 2001. Published with permission from APCO's Public Safety Communications/APCO Bulletin magazine.

SUMMARY

This chapter has reviewed many of the ideas and thoughts in this book to help you prepare for a possible terrorist attack—in your area, or at your center. This entire book is a start, a place to look at all aspects of terrorism preparedness. Now is your chance to go forward and implement ways to secure your facility and staff. The hope of all of the authors in this book is that we never suffer another tragedy as that which occurred on September 11th, 2001. We know that communications personnel will handle whatever crisis that faces their communities and their agency with the same professionalism that occurred on that day and every day in centers across the country.

APPENDIX A

Public Safety Precautions/Actions

GENERAL FACTORS

INDICATORS OF BIOLOGICAL AGENT

Biological agents have the potential to be more lethal than chemical agents and are primarily deployed through aerosol spray or introduction into a water system. There have been two documented Biological Weapons (BW) attacks in the United States. One occurred in 1984 when followers of the Rajineesh Bagwhan produced and dispensed salmonella bacteria in Oregon. In that case, the perpetrators spread the agent via restaurant salad bars. The other occurred in the latter part of 2001, when anthrax-laced letters were mailed through the United States postal system. At the time of this book's publication, the perpetrator of this crime was unknown.

The following are indications of a biological terrorism event.

- A single, definitively diagnosed or strongly suspected case of an illness due to a potential bioterrorist agent occurring in a patient with no history suggesting another explanation for the illness

- A cluster of patients presenting with a similar disease with unusually high morbidity or mortality without an obvious etiology or explanation

- An unexplained increase in the incidence of a common syndrome above seasonally expected levels or with higher than expected morbidity and mortality

BIOLOGICAL WEAPONS FIRST RESPONDER CONCERNS

Treat all incidents involving biological agents as intentional hazardous materials situations. In all cases, safely isolate and deny entry and make appropriate notifications. In addition, whenever it is believed that a biological agent has been released, assume that all personnel and property has been contaminated.

The most practical method of initiating a biological attack is by dispersal of aerosol particles. Biological agents may be able to enter the body through the inhalation, ingestion, or direct contact with skin or membranes. Unlike chemical agents, exposure to biological agents may not be immediately apparent, with casualties occurring hours, days, or weeks after exposure. In a silent release scenario, the first indication of a biological attack may occur after a number of unusual illnesses begin to appear in local hospital emergency departments or through an increase of calls to communications centers. Without advance warning, first responders may not recognize the existence of a biological attack. Additionally, first responders should immediately

- Don PPE if available.

- Request specialized resources. These resources include public health officials from the county and state departments of health. Experts agencies such as the CDC, the MMRS, NMRT-WMD, and the USAMRI are also needed to identify the nature of the biological agent.

- Take measures to prevent an epidemic (pending identification of the agent). These measures include isolation, quarantine, and restriction of personnel movement based on local, state, and federal laws and regulations. These procedures apply to both victims and first responding personnel. Identify the source of contamination and designate zones of operation (i.e., hot, warm, and cold zones). Consider weather effects during zone designation. If large numbers of exposures are involved, quarantine may be necessary with all victims being treated on-site. If a small number of persons are exposed, they should be decontaminated and transported to a hospital capable of isolating the patients.

- Initiate protective actions (i.e., evacuation or in-place protection).

- Consider the impact of weather conditions. The impact of biological agents are affected by weather conditions. Accordingly, detailed and accurate assessments of weather conditions and forecasts are critical elements in the tactical management of biological emergencies. Weather effects to consider include

  - Sunlight. Ultraviolet light found in sunlight helps kill biological agents.

  - Temperature. Temperatures above 100°F begin killing biological agents. Freezing temperatures can render biological agents dormant.

  - Temperature gradient. Elevation influences temperature. For each ten meters from ground level, there is a different temperature known as the temperature gradient. This factor causes biological agents to hold close to the ground.

  - Wind. Wind aids the dispersal and spread of biological agents. Wind direction and speed influence the resulting plume and must be considered when setting up zones of operation and making evacuation decisions.

  - Precipitation. Precipitation can influence agent dispersal and the spread of contaminated areas (e.g., run-off). In biological situations, the quantity of rain can either kill or stimulate the growth of individual agents.

- Humidity. Higher humidity levels cause the pores on human skin to open up, aiding the absorption of agents.

INITIAL ACTIONS BY FIRST RESPONDERS

In cases of actual release of biological agents, first responding units must immediately take steps to protect themselves. First responders suspecting a biological release must remain calm, don PPE, aid from a safe vantage point, reassure victims that assistance is on the way, and wait for properly equipped help at a safe location (upwind, uphill, upstream). Communications centers can assist by prompting first responder actions and offering precaution information to help police, fire, and medical personnel respond safely. This involves safely isolating, denying entry, and making notifications. The following checklist summarizes the essential ingredients of an initial biological notification:

- Observed biological indicators

- Wind direction and weather conditions at scene

- Plume direction (direction of cloud or vapor travel)

- Orientation of victims (direction, position, and pattern)

- Number of apparent victims

- Type of injuries and/or symptoms presented

- Witnesses statements or observations

- Nature of biological agents (if known) from detection equipment or monitors

- Exact location of reporting unit

- Suggested safe access route and staging area

INITIAL ACTIONS BY DISPATCH PERSONNEL

Dispatch personnel play a key role in mobilizing proper response and support to a biological incident. Public-safety dispatchers (both law enforcement and fire services) are vital elements in recognizing and assessing biological events. Dispatchers must be aware of potential target locations and the indicators of possible criminal or terrorist activity involving biological agents. Dispatchers must know the signs of exposure to biological agents and recognize unusual trends or patterns of activity indicative of a possible biological incident. Dispatchers must also be able to discern and solicit critical information regarding threats and biological indicators encountered by field personnel.

DECONTAMINATION

Biological incidents may potentially involve civilians, law enforcement, fire services, and medical personnel that have been exposed to potentially lethal agents. Prompt, safe, and effective decontamination procedures are essential to protect exposed persons, equipment, and the environment from the harmful effects of these agents.

Decontamination is the process used to reduce the hazards of biological agents to safe levels. It minimizes the uncontrolled transfer of contaminants from the hazard site to clean areas, and should be performed any time contamination with a biological agent or hazardous material is suspected.

During decontamination operations, the safety of emergency response personnel is the first and most important consideration. Proper use of PPE such as SCBA reduces hazards to response personnel.

The risk of secondary contamination to rescue personnel, medical personnel on the scene and at the hospital, and transport vehicles and equipment must be adequately assessed and protected against to avoid spreading the incident. Any contamination of the skin must be decontaminated immediately.

Hazardous materials teams must establish standard decontamination procedures for a range of biological incidents. These procedures should include provisions for selecting and establishing a decontamination site as well as specific operational protocols. All personnel assigned to these teams shall be thoroughly trained to safely and effectively carry out their responsibilities. Specific decontamination protocols must retain the flexibility to respond to a range of hazards or conditions at the incident scene or decontamination site.

Decontamination efforts can be hampered by several factors. Often, the very properties that make an agent dangerous also make it difficult to decontaminate. The larger the area affected, the more difficult the decontamination process will be, and the longer a contaminant is in contact with an object, the longer it will take to clean it up. Different protective gear offers different levels of protection, and presents different decontamination needs. Other factors to consider in the decontamination process include

- Prevention of further contamination
- Minimization of contact with potential contaminants to keep the incident from escalating
- Physical and chemical properties of the agent
- Amount and location of contamination
- Contact time and temperature
- Level of protection and work function

PATIENT MANAGEMENT

The management of patients following suspected or confirmed bioterrorism events must be well organized. Strong leadership and effective communication are paramount. Remember that communications centers will be inundated with questions about the event and exposure and requests for family information and treatment. Communications centers should be involved in planning of all phases of a biological event.

ISOLATION PRECAUTIONS

Agents of bioterrorism are generally not transmitted from person to person—re-aerosolization of these agents is unlikely. All patients in health-care settings, including symptomatic patients with suspected or confirmed bioterrorism-related illnesses, should be managed utilizing standard precautions. Standard precautions are designed to reduce transmission from both recognized and unrecognized sources of infection in health-care facilities, and are recommended for all patients receiving care, regardless of their diagnosis or presumed infection status. For certain diseases or syndromes (e.g., smallpox and pneumonic plague), additional precautions may be needed to reduce the likelihood for transmission.

Standard precautions prevent direct contact with all body fluids (including blood), secretions, excretions, non intact skin (including rashes), and mucous membranes. Standard precautions routinely practiced by health-care providers include

- Washing hands after touching blood, body fluids, excretions, secretions, or items contaminated with such body fluids, whether or not gloves are worn. Hands are washed immediately after gloves are removed, between patient contacts, and as appropriate to avoid transfer of microorganisms to other patients and the environment. Either plain or antimicrobial soaps may be used according to facility policy.

- Wearing clean, nonsterile gloves when touching blood, body fluids, excretions, secretions, or items contaminated with such body fluids. Clean gloves are put on just before touching mucous membranes and nonintact skin. Gloves are changed between tasks and between procedures on the same patient if contact occurs with contaminated material. Hands are washed promptly after removing gloves and before leaving a patient care area.

- Wearing a mask and eye protection (or face shield) to protect mucous membranes of the eyes, nose, and mouth while performing procedures and patient care activities that may cause splashes of blood, body fluids, excretions, or secretions.

- Wearing a gown to protect skin and prevent soiling of clothing during procedures and patient-care activities that are likely to generate splashes or sprays of blood, body fluids, excretions, or secretions. Selection of gowns and gown materials should be suitable for the activity and amount of body fluid likely to be encountered. Soiled gowns are removed promptly and hands are washed to avoid transfer of microorganisms to other patients and environments.

PATIENT PLACEMENT

In small-scale events, routine patient-placement and infection-control practices should be followed. When the number of patients arriving at a health-care facility is too large to allow routine triage and isolation strategies (if required), it is necessary to apply practical alternatives. These may include grouping patients who present with similar syndromes into a designated section of a clinic or emergency department, a designated ward or floor of a facility, or a response center at a separate building. Communications centers may be asked to assist in arranging patient placement and resources to do this. Communications centers should be prepared to assist in any situation. They are as much a vital link in patient care as the first responders.

PATIENT TRANSPORT

Most infections associated with bioterrorism agents cannot be transmitted from patient to patient. In general, however, the transport and movement of patients with bioterrorism-related infections should be limited to movement that is essential to provide patient care, thus reducing the opportunities for transmission of microorganisms within health-care facilities.

DISCHARGE MANAGEMENT

Ideally, patients with bioterrorism-related infections will not be discharged from the facility until they are deemed noninfectious. However, consideration should be given to developing home-care instructions in the event that large numbers of persons exposed may preclude admission of all infected patients. Depending on the exposure and illness, home-care instructions may include recommendations for the use of appropriate barrier precautions, hand washing, waste management, and cleaning and disinfection of the environment and patient-care items. If this should occur, communications centers can expect an increase in the number of calls from family, neighbors, and others who may not understand the need for bed space in a large scale exposure. Keep in touch with local hospitals and health departments for updated information about the exposure so you can pass along proper information to the calling public.

PATIENTS AND ENVIRONMENT

The need for decontamination depends on the suspected exposure and in most cases will not be necessary. The goal of decontamination after a potential exposure to a bioterrorism agent is to reduce the extent of external contamination of the patient and contain the contamination to prevent further spread. Decontamination should only be considered in instances of gross contamination. Decisions regarding the need for decontamination should be made in consultation with state and local DOHs. It may be necessary to decontaminate exposed individuals prior to receiving them in the health-care facility to ensure the safety of patients and staff while providing care.

Depending on the agent, the likelihood for re-aerosolization, and the risk associated with cutaneous exposure, clothing of exposed persons may need to be removed. After removal of contaminated clothing, patients should be instructed (or assisted, if necessary) to immediately shower with soap and water. Potentially harmful practices, such as bathing patients with bleach solutions, are unnecessary and should be avoided. Clean water, saline solution, or commercial ophthalmic solutions are recommended for rinsing eyes. If indicated, after removal at the decontamination site, patient clothing should be handled only by personnel wearing appropriate PPE and be placed in an impervious bag to prevent further environmental contamination.

Be aware that the FBI may require collection of exposed clothing and other potential evidence for submission to FBI or DoD laboratories to assist in exposure investigations.

PROPHYLAXIS AND POST-EXPOSURE IMMUNIZATION

Recommendations for prophylaxis are subject to change. Up-to-date recommendations should be obtained in consultation with local and state DOHs and the CDC. Facilities should ensure that policies are in place to identify and manage health-care workers exposed to infectious patients. In general, maintenance of accurate occupational health records will facilitate identification, contact, assessment, and delivery of post-exposure care to potentially exposed health-care workers. Communications centers can use their database to provide address information for post-exposure public immunization efforts. Reverse 9-1-1 can also be used to provide the public with proper information concerning post-exposure care.

TRIAGE AND MANAGEMENT OF LARGE SCALE EXPOSURES AND SUSPECTED EXPOSURES

Each health-care facility, with the involvement of the IC, administration, building engineering staff, the emergency department, laboratory directors, and nursing directors should clarify in advance how they will best be able to deliver care in the event of a large-scale exposure. Facility needs will vary with the size of the population served and the proximity to other health-care facilities and external assistance. Triage and management planning for large-scale events should include

- Establishing networks of communication and lines of authority required to coordinate on-site care

- Planning for cancellation of nonemergency services and procedures

- Identifying sources able to supply vaccines, immune globulin, antibiotics, and botulinum antitoxin (with assistance from local and state DOHs)

- Planning for the efficient evaluation and discharge of patients

- Developing discharge instructions for patients determined to be noncontagious or in need of additional on-site care, including details regarding if and when they should return for care and whether they should seek medical follow-up

PSYCHOLOGICAL ASPECTS OF WMD EVENTS

Following a WMD-related event, fear and panic can be expected from patients, health-care providers, and the general public. Psychological responses following a WMD event may include horror, anger, panic, unrealistic concerns about infection, fear of contamination, paranoia, social isolation, or demoralization.

Consider the following to address patient and general-public fears:

- Minimize panic by clearly explaining risks, offering careful but rapid medical evaluation and treatment, and avoiding unnecessary isolation or quarantine.

- Treat anxiety in unexposed persons who are experiencing somatic symptoms. Do this with reassurance, or diazepam-like anxiolytics as indicated for acute relief of those who do not respond to reassurance.

Consider the following to address first-responder health fears:

- Provide WMD-readiness education, including frank discussions of potential risks and plans for protecting health-care providers.

- Invite active, voluntary involvement in the WMD-readiness planning process.

- Encourage participation in disaster drills.

The following Job Aid is from FEMA's Independent Study Course IS-15: *Special Events Contingency Planning for Public Safety Agencies.*

JOB AID: FIRST RESPONSE TO A TERRORIST INCIDENT

The following are some guidelines developed by the Pennsylvania Emergency Management Agency for responders on the scene of a terrorist incident.

THE TEN "ATES"

1. **Evacuate** the area as quickly and safely as possible.

2. **Isolate** the site to restrict access by all personnel.

3. **Hesitate**. Fools rush in. Do not be one. Do not enter the scene until you:

4. **Evaluate** the situation and your potential response actions.

5. **Communicate** your conclusions and call for assistance as necessary (because of bomb danger, use NO radios or cellular phones closer than 300 yards). Notify hospitals as appropriate.

6. **Infiltrate**. Go in carefully, and only when it is time to do so.

7. **Procrastinate**. Take no action until it is as safe as possible and necessary.

8. **Investigate**. Remember, this is a crime scene. Do not exceed your authority but support and assist the investigation as appropriate.

9. **Cooperate** with other responders (teamwork!) and with those in charge.

10. **Decontaminate** and clean up carefully to avoid accidental removal of evidence and to avoid endangering others.

GOLDEN RULE FOR FIRST RESPONDERS

Do not touch anything at a crime scene or remove anything from a crime scene unless 1) it is absolutely necessary for the performance of your official duties, or 2) it is done in concurrence with the appropriate law enforcement personnel.

REMEMBER

THIS IS A CRIME SCENE. It is the scene of a deliberately violent and lethal act. THERE MAY BE MORE ACTS.

WORDS TO LIVE BY

- Do not create more casualties "rescuing" the dead.

- Life safety—of 1) responders and 2) victims—is first priority.

- Second priority is preservations of evidence.

- Examine victims for injuries and weapons. A perpetrator may have been injured, too.

CHEMICAL INCIDENT

- Approach from upwind if possible.

- Use PPE.

- Stay clear of spills, vapors, fumes, and smoke.

- Exclusion zone: 1,500 feet.

- Use fog streams instead of solid streams where possible to preserve evidence.

- Contain runoff where possible.

- Cover all entries with charged lines to prevent flare-ups.

BOMB

- *Assume there are more!* Responders may be terrorist targets, too.

- Establish 300-yard exclusion zone.

- Do *not* use radios or cellular phones within 300 yards of the site.

- Remove the injured as quickly and carefully as possible; leave the dead for coroners.

BIOLOGICAL/NERVE AGENTS

- Watch for numbers of people or animals exhibiting similar symptoms of illness.

- Watch for human or animal remains with no apparent trauma.

- *IMMEDIATELY* don respirator and leave area if situation is suspicious.

NUCLEAR/RADIOLOGICAL INCIDENT

- This is not detectable without monitoring equipment.

- Distance is best immediate protection; enforce bomb exclusion zone.

- Remember: "Time, Distance, Shielding."

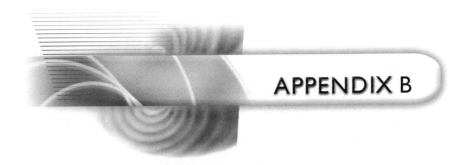

APPENDIX B

Anthrax Threat Advisory

October 12, 2001, Center for Disease Control (CDC)

HANDLING ANTHRAX AND OTHER BIOLOGICAL AGENT THREATS

Many facilities in communities around the country have received anthrax threat letters. Most were empty envelopes; some have contained powdery substances. The purpose of these guidelines is to recommend procedures for handling such incidents.

DO NOT PANIC

1. Anthrax organisms can cause infection in the skin, gastrointestinal system, or the lungs. To do so, the organism must be rubbed into abraded skin, swallowed, or inhaled as a fine, aerosolized mist. Disease can be prevented after exposure to the anthrax spores by early treatment with the appropriate antibiotics. Anthrax is not spread from person to person.

2. For anthrax to be effective as a covert agent, it must be aerosolized into very small particles. This requires a great deal of technical skill and special equipment. If these small particles are inhaled, life-threatening lung infection can occur, but prompt recognition and treatment is effective.

SUSPICIOUS UNOPENED LETTER OR PACKAGE MARKED WITH THREATENING MESSAGE SUCH AS "ANTHRAX"

1. *Do not shake or empty* the contents of any suspicious envelope or package.

2. *PLACE the envelope or package in a plastic bag* or some other type of container to prevent leakage of contents.

3. If you do not have any container, then *COVER the envelope or package* with anything (e.g., clothing, paper, trash can, etc.) and do not remove this cover.

4. *LEAVE the room and CLOSE the door*, or section off the area to prevent others from entering. (i.e., keep others away).

5. *WASH your hands with soap and water* to prevent spreading any powder to your face.

6. If you are at home, *report the incident to local police.* If you are at work, report the incident to local police *and* notify your building security official or an available supervisor.

7. *List all of the people who were in the room* or area when the suspicious letter or package was recognized. Give this list to both the local public health authorities and law enforcement officials for follow-up investigations and advice.

If the powder spills out of the envelope and onto a surface,

1. *Do not try to clean up the powder.* Cover the spilled contents immediately with anything (e.g., clothing, paper, trash can, etc.) and do not remove this cover!

2. *Leave the room and close the door* or section off the area to prevent others from entering (i.e., keep others away).

3. *Wash your hands with soap and water* to prevent spreading any powder to your face.

4. If you are at home, *report the incident to local police.* If you are at work, report the incident to local police *and* notify your building security official or an available supervisor.

5. *Remove heavily contaminated clothing* as soon as possible and place in a plastic bag or some other container that can be sealed. This clothing bag should be given to the emergency responders for proper handling.

6. *Shower with soap and water* as soon as possible. *Do not use bleach or other disinfectant on your skin.*

7. If possible, especially those who had actual contact with the powder. Give this list to both the local public health authorities so that proper instructions can be given for medical follow-up and to law enforcement officials for further investigation.

POSSIBLE ROOM CONTAMINATION BY AEROSOLIZATION

(For example, small device triggered, warning that air handling system is contaminated, or warning of a biological agent released in a public space.)

1. *Turn off local fans or ventilation units* in the area.

2. *Leave area immediately.*

3. *Close the door or section off the area* to prevent others from entering (i.e., keep others away).

4. If you are at home, *dial 9-1-1* to report the incident to local police and the local FBI field office. If you are at work, dial 9-1-1 to report the incident to local

police and the local FBI field office *and* notify your building security official or an available supervisor.

5. *Shut down air handling system* in the building, if possible.

6. If possible, *list all people who were in the room or area.* Give this list to the local public health authorities so that proper instructions can be given for medical follow-up and to law enforcement officials for further investigation.

HOW TO IDENTIFY SUSPICIOUS PACKAGES AND LETTERS

Some characteristics of suspicious packages and letters include the following:

- Excessive postage
- Handwritten or poorly typed addresses
- Incorrect titles
- Title but no name
- Misspellings of common words
- Oily stains, discolorations, or odor
- No return address
- Excessive weight
- Lopsided or uneven envelope
- Protruding wires or aluminum foil
- Excessive security material such as masking tape, string, etc.
- Visual distractions
- Ticking sound
- Marked with restrictive endorsements, such as "Personal" or "Confidential"
- Shows a city or state in the postmark that does not match the return address

Indicators of Possible Agent Usage

INDICATORS OF POSSIBLE CHEMICAL WARFARE (CW) AGENT USAGE

Contact the fire department or emergency management office for assistance during your planning phase as it relates to chemical, biological, and nuclear agents.

UNUSUAL DEAD OR DYING ANIMALS

- Lack of insects in the air

UNEXPLAINED CASUALTIES

- Multiple victims

- Serious illnesses

- Nausea, disorientation, difficulty breathing, convulsions

- Definite casualty patterns

UNUSUAL LIQUID, SPRAY, OR VAPOR

- Droplets, oily film

- Unexplained odor

- Low-lying clouds or fog unrelated to weather

SUSPICIOUS DEVICES/PACKAGES

- Unusual metal debris

- Abandoned spray devices

- Unexplained munitions

INDICATORS OF POSSIBLE BIOLOGICAL WEAPON (BW) USAGE

UNUSUAL DEAD OR DYING ANIMALS

- Sick or dying animals

UNUSUAL CASUALTIES

- Unusual illness for region or area

- Definite pattern inconsistent with natural disease

UNUSUAL LIQUID, SPRAY, OR VAPOR

- Spraying

- Suspicious devices or packages

IF RELEASE IS SUSPECTED

- Remain calm.

- Don PPE.

- Establish control of the scene.

- From a safe vantage point, reassure victims assistance is on the way.

- Request properly equipped help at a safe location (upwind, uphill, upstream).

- Safely isolate and deny entry to the affected area.

NOTIFICATION ESSENTIALS

- Observed NBC indicators

- Wind direction and weather conditions at scene

- Plume direction (direction of cloud or vapor travel)

- Orientation of victims (direction, position, pattern)

- Number of apparent victims

- Type of injuries, symptoms presented

- Witnesses statements or observations

- Nature of NBC agents (if known) from detection equipment or monitors

- Suggested safe access route and staging area

INCIDENT OBJECTIVES

- Secure a perimeter and designate zones of operation (hot, warm, cold).

- Control and identify agent.

- Rescue, decontaminate, triage, treat, and transport affected people.

- Move uninvolved crowds or persons to safe zones.

- Stabilize the incident.

- Avoid secondary contamination.

- Secure evidence and crime scene.

- Protect against secondary attack.

Upon the recognition of the above indicators, refer to your standard operating procedures for local, state, and federal agencies to contact. This contact list should be established prior to a WMD event. Suggested agencies are local and state emergency management, law enforcement, fire services and hazmat; federal agencies are FEMA, FBI, and other related support and response agencies.

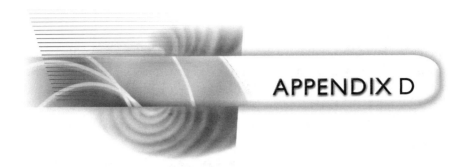

APPENDIX D

Complete Agent Description

ANTHRAX

DESCRIPTION OF AGENT

Anthrax is a highly lethal infection caused by the Gram-positive bacterium *Bacillus anthracis*. In naturally acquired cases, organisms usually enter through skin wounds (causing a localized infection), but they may be inhaled or ingested. Intentional release by belligerents or terrorist groups would presumably involve the aerosol route, as the spore form of the *bacillus* is quite stable and possesses characteristics ideal for the generation of aerosols.

SIGNS AND SYMPTOMS

The incubation period for inhalational anthrax is one to six days. Fever, malaise, fatigue, cough, and mild chest discomfort are rapidly followed by severe respiratory distress with dyspnea, diaphoresis, stridor, and cyanosis. Shock and death occur within twenty-four to thirty-six hours of the onset of severe symptoms. In cases of cutaneous anthrax, a papule develops, then vesicles, finally developing into a black eschar surrounded by moderate to severe edema. The lesions are usually painless. Without treatment, cutaneous anthrax may progress to septicemia and death, with a case-fatality rate of twenty percent. With treatment, fatalities are rare.

DIAGNOSIS

Physical findings are typically nonspecific in inhalational cases, with initial complaints of malaise, fever, headache, and possibly substemal chest pain. A widened mediastinum is sometimes seen on an x-ray late in the course of the illness and correlates with a pathologic finding of hemorrhagic mediastinitis, the classic presentation of inhalational anthrax. The bacterium may be detected by a Gram stain of blood and by a blood culture late in the course of the illness.

TREATMENT

Although usually ineffective in inhalational cases once symptoms are present, antibiotic treatment with high-dose penicillin, ciprofloxacin, or doxycycline should nonetheless be administered. Although typically sensitive to penicillin, resistant isolates are readily produced in the laboratory. For this reason, in the case of an intentional release and in the absence of antibiotic sensitivity data, treatment should be initiated with 117V ciprofloxacin (400 mg q 8–12 hrs) or IV doxycycline (200 mg initially, followed by 100 mg q 12 hrs). Supportive therapy may be necessary.

PROPHYLAXIS

A licensed vaccine is available for use for those at risk of exposure. Vaccination is undertaken at zero, two, and four weeks (initial series), followed by booster doses at six, twelve, and eighteen months, and then yearly. Oral ciprofloxacin (500 mg po bid) or doxycycline (100 mg po bid) are useful in cases of known or imminent exposure. Following confirmed exposure, all unimmunized individuals should receive three 0.5 ml SQ doses of vaccine over thirty days, while those vaccinated with less than three doses prior to exposure should receive an immediate 0.5 ml booster. Anyone vaccinated with the initial three-dose series in the previous six months does not require boosters. All exposed personnel should continue antibiotic therapy for four weeks. If the vaccine is unavailable, antibiotics may be continued beyond four weeks and should be withdrawn only under medical supervision.

DECONTAMINATION AND ISOLATION

Drainage and secretion precautions should be practiced. Anthrax is not known to be transmitted via the aerosol route from person to person. Following invasive procedures or autopsy, instruments and surfaces should be thoroughly disinfected with a sporicidal agent (high-level disinfectants such as iodine or 0.5 percent sodium hypochlorite).

OUTBREAK CONTROL

Anthrax spores may survive in the environment for many years, secondary aerosolization of such spores (such as by pedestrian movement or vehicular traffic) generally presents no problem for humans. The carcasses of animals that die in such an environment should be burned, and animals subsequently introduced into such an environment should be vaccinated. Meat, hides, and carcasses of animals in affected areas should not be consumed or handled by untrained and/or unvaccinated personnel.

BRUCELLOSIS

DESCRIPTION OF AGENT

Human brucellosis is an infection caused by one of four species of Gram-negative coccobacilli of the genus *Brucella*. *B. abortus* is normally a pathogen of cattle, while *B. melitensis, B. suis, and B. canis* are pathogens of goats, pigs, and dogs, respectively. Organisms are acquired by humans via the oral route through the ingestion of unpasteurized milk and cheese, via inhalation of aerosols generated on farms and in slaughterhouses, or via inoculation of skin lesions in persons with close animal contact. Intentional exposure by belligerents would likely involve aerosolization, but it could involve contamination of foodstuffs.

SIGNS AND SYMPTOMS

The incubation period is quite variable, with symptoms often requiring months to appear. This marked variability would appear to somewhat temper the use of *brucellae* as weapons. Symptoms of acute and subacute brucellosis are nonspecific and consist of irregular fever, headache, profound weakness and fatigue, chills and sweating, and generalized arthralgias and myalgias. Depression and mental status changes are noteworthy. Osteoarticular complications, particularly involving the axial skeleton sacroiliitisvertebral osteomyelitis), are common. Fatalities are uncommon, even in the absence of therapy.

DIAGNOSIS

Naturally occurring cases may often be suspected based on a history of close animal contact or consumption of implicated foodstuffs. *Brucellae* may be isolated from standard blood cultures but require a prolonged period of incubation; cultures should thus be maintained for six weeks if brucellosis is suspected. Bone marrow cultures yield the diagnosis in a higher percentage of cases than do peripheral blood cultures. A serum agglutination test is available and often helpful.

TREATMENT

Doxycycline (100 mg po bid) plus rifampin (600-900 mg po qd) administered for six weeks is the regimen of choice for uncomplicated brucellosis. Doxycycline plus streptomycin, TMP/SMX plus gentamicin, and ofloxacin plus rifampin are acceptable alternative regimens.

PROPHYLAXIS

Avoidance of nonpasteurized milk products and appropriate veterinary vaccination practices are sufficient to prevent most naturally occurring brucellosis. Persons inadvertently exposed to veterinary vaccine strains of brucella have been successfully prophylaxed with doxycycline plus rifampin for ten days. No human brucellosis vaccine is available in the western world.

DECONTAMINATION AND ISOLATION

Drainage and secretion precautions should be practiced in patients who have open skin lesions; otherwise no evidence of person-to-person transmission of brucellosis exists. Animal remains should be handled according to universal precautions and disposed of properly. Surfaces contaminated with *brucella* aerosols may be decontaminated by standard means (0.5 percent hypochlorite).

OUTBREAK CONTROL

In the event of an intentional release of *brucella* organisms, it is possible that livestock will become infected. Thus, animal products in such an environment should be pasteurized, boiled, or thoroughly cooked prior to consumption. Proper treatment of water, by boiling or iodination, would also be important in an area subjected to intentional contamination with *brucella* aerosols.

PLAGUE

DESCRIPTION OF AGENT

Plague is an infectious disease caused by the Gram-negative, bipolar staining bacterium *Yersinia pestis*. Naturally occurring plague is most often acquired by the bite of a flea that had previously fed on infected rodents. In such cases, plague classically presents as a localized abscess with secondary formation of very large, fluctuant regional lymph nodes known as buboes (bubonic plague). Plague may also be transmitted via aerosols and by inhalation of sputum droplets from coughing patients. In such instances, a primary pneumonic form may develop and, in the absence of prompt therapy, progress rapidly to death within two to three days. Intentional release by belligerents or terrorist groups would presumably involve aerosolization, but could also involve the release of infected fleas. Plague may be considered a lethal agent.

SIGNS AND SYMPTOMS

Pneumonic plague has an incubation period of two to three days and begins with high fever, chills, headache, hemoptysis, and toxemia, progressing rapidly to dyspnea, stridor, and cyanosis. Death results from respiratory failure, circulatory collapse, and bleeding diatheses. Bubonic plague has an incubation period of two to ten days and presents with malaise, high fever, and tender lymph nodes (buboes). Bubonic plague may progress spontaneously to the septicemic form, with spread to the central nervous system, lungs, and elsewhere.

DIAGNOSIS

To facilitate prompt therapy, plague must be suspected clinically. A presumptive diagnosis may also be made by a Gram or Wayson stain of lymph node aspirates, sputum, or Cerebral Spinal Fluid (CSF). The plague bacillus may be readily cultured from aspirates of buboes or from the blood of septicemic patients.

TREATMENT

Early administration of antibiotics is quite effective, but must be started within twenty-four hours of onset of symptoms in pneumonic plague. The treatment of choice is streptomycin (30 mg/kg/day IN4 in two divided doses for 10 days) or gentamicin (2 mg/kg, then 1.0–1.5 mg/kg q 8 hrs for 10 days). Intravenous doxycycline (200 mg, then 100 mg q 12 hrs for 10–14 days) is also effective; chloramphenicol should be added in cases of plague meningitis. Supportive therapy for pneumonic and septicemic forms is typically required

PROPHYLAXIS

A licensed, killed vaccine is available. The primary vaccination series consists of a 1.0 ml IM dose initially, followed by 0.2 ml doses at one to three months, and three to six months. Booster doses are given at six, twelve, and eighteen months and then every one to two years. As this vaccine appears to offer no protection against aerosol exposure in animal experiments, victims of a suspected attack with aerosolized plague or those who have respiratory contacts with coughing patients should be given doxycycline (100 mg po bid for 7 days or the duration of exposure, whichever is longer).

DECONTAMINATION AND ISOLATION

Drainage and secretion precautions should be employed in managing patients with bubonic plague; such precautions should be maintained until the patient has received antibiotic therapy for forty-eight hours and has demonstrated a favorable response to such therapy. Care must be taken when handling or aspirating buboes to avoid aerosolizing infectious material. Strict isolation is necessary for patients with pneumonic plague.

OUTBREAK CONTROL

In the event of the intentional release of plague into an area, it is possible that local fleas and rodents could become infected, thereby initiating a cycle of enzootic and endemic disease. Such a possibility would appear more likely in the face of a break-down in public-health measures (such as vector and rodent control), which might accompany armed conflict. Care should be taken to rid patients and contacts of fleas utilizing a suitable insecticide. Flea and rodent control measures should be instituted in areas where plague cases have been reported.

TULAREMIA

DESCRIPTION OF AGENT

Tularemia is an infection caused by the Gram-negative coccobacillus *Francisella tularensis*. Two biogroups are known. Biogroup *tularensis*, also known as type A, is the more virulent form and is endemic in much of North America. Naturally acquired tularemia is contracted through the bites of certain insects (notably ticks and deerflies) or via contact with infected rabbits, muskrats, and squirrels. Intentional release by belligerents would presumably involve aerosolization of living organisms. Although naturally acquired tularemia has a case fatality rate of approximately 5 percent, the pneumonic form of the disease, which would predominate in the setting of intentional release, would likely have a greater mortality rate.

SIGNS AND SYMPTOMS

Naturally acquired tularemia frequently has an ulceroglandular presentation, although a significant minority of cases involve the typhoidal or pneumonic forms. The incubation period averages three to five days, but varies widely. Use of tularemia as a weapon would likely lead to a preponderance of pneumonic and typhoidal cases, and large aerosolized innocula would be expected to shorten the incubation period. Ulceroglandular disease involves a necrotic, tender ulcer at the site of inoculation, accompanied by tender, enlarged regional lymph nodes. Fever, chills, headache, and malaise often accompany these findings. Typhoidal and pneumonic forms often involve significant cough, abdominal pain, substernal discomfort, and prostration in addition to prolonged fever, chills, and headache.

DIAGNOSIS

Prompt diagnosis relies on clinical suspicion. Routine laboratory tests are rarely helpful, and *F. tularensis* does not typically grow in standard blood cultures, although special media are available for the culturing (under BL-3 containment conditions) of blood, sputum, lymph node material, and wound exudates if the diagnosis is suspected. Serology is available to confirm the diagnosis in suspected cases.

TREATMENT

Streptomycin (7.5–15 mg/kg IM Q 12 hrs for 7–14 days) is the drug of choice for all forms of tularemia. Gentamicin (3–5 mg/kg/d q 8–12 hrs for 7–14 days) is an acceptable alternative. Relapses are more common with tetracycline therapy (500 mg po q 6 hrs for 14 days), although this alternative may be employed in patients who cannot tolerate aminoglycosides.

PROPHYLAXIS

A live, attenuated vaccine is available as an investigational product through USAM-REED (Fort Detrick, Md.). It may be given to those individuals, such as laboratory workers, at high risk of exposure. A single dose is administered by scarification. Intramuscular streptomycin will prevent disease following documented exposure, but is not recommended following tick bites or animal contact.

DECONTAMINATION AND ISOLATION

Tularemia is not transmitted person to person via the aerosol route, and infected persons should be managed with secretion and drainage precautions. Heat and common disinfectants (such as 0.5 percent hypochlorite) will readily kill *F. tularensis* organisms.

OUTBREAK CONTROL

Following intentional release of *F. tularensis* in a given area, it is possible that local fauna, especially rabbits and squirrels, will acquire the disease, setting up an enzootic mammal-arthropod cycle. Persons entering a contaminated area should avoid skinning and eating meat from such animals. Water supplies and grain in contaminated areas might likewise become contaminated and should be boiled or cooked before consumption. Organisms that contaminate soils are unlikely to survive for significant periods of time and present little hazard.

SMALLPOX

DESCRIPTION OF AGENT

Smallpox is an infection caused by the Variola virus, a member of the chordopoxvirus family. Naturally occurring smallpox has been eradicated from the globe, with the last case occurring in Somalia in 1977. Repositories of the virus are known to exist in only two laboratories worldwide. Monkeypox, cowpox, and vaccinia are closely related viruses that might lend themselves to genetic manipulation and the subsequent production of smallpox-like disease.

SIGNS AND SYMPTOMS

The incubation period of smallpox is about twelve days. Clinical manifestations begin acutely with a prodromal period involving malaise, fevers, rigors, vomiting, headache, and backache. After two to four days, skin lesions appear and progress uniformly from macules to papules to vesicles and pustules. Lesions progress centrifugally and scab in one to two weeks. In unvaccinated individuals, variola major, the classical form of the disease, is fatal in approximately 30 percent of cases.

DIAGNOSIS

In its full-blown form as typically seen in unimmunized individuals, smallpox is readily diagnosed on clinical grounds. Differentiation from other vesicular exanthems such as varicella and erythema multiforme might be difficult, however, in cases of variola minor or in disease modified by prior vaccination. Electron microscopy can readily differentiate variola virus from varicella, but not from vaccinia and monkeypox when performed on lesion scrapings. The virus can be grown in chorioallantoic membrane culture.

TREATMENT

Supportive care is the mainstay of smallpox therapy. No specific antiviral therapy exists.

PROPHYLAXIS

A licensed, live vaccinia virus vaccine is available and is administered via a bifurcated needle using a multiple puncture technique (scarification). Given the eradication of smallpox, the vaccine would only be indicated in laboratory settings or where biological warfare was a distinct possibility. Vaccination is probably protective for at least three years. Exposed persons may be managed with prompt vaccination. Vaccinia Immune Globulin (VIG), given IN4 at a dose of 0.6 ml/kg, may prove a useful adjunct to vaccination, although its precise role is unclear.

DECONTAMINATION

Given the extreme public-health implications of smallpox reintroduction, patients should be placed in strict isolation pending review by national health authorities. All material used in patient care or in contact with smallpox patients should be autoclaved, boiled, or burned.

OUTBREAK CONTROL

Smallpox has considerable potential for person-to-person spread. Thus, all contacts of infectious cases should be quarantined for sixteen to seventeen days following exposure and given prophylaxis as indicated. Animals are not susceptible to smallpox.

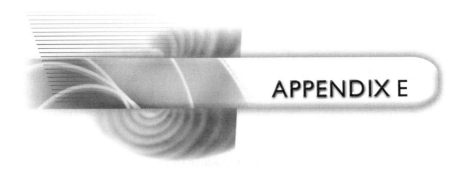

Internet Resources for Terrorism/ Disaster Planning

For those agencies with access to the Internet, there is a wealth of information available to assist you in the development and research of your plan. Listed below are a few links related to terrorism.

COMMUNICATIONS CENTER LINKS

Association of Public Safety Communications Officials (APCO)

www.APCO911.org

National Emergency Number Association (NENA)

www.nena.org

TERRORISM LINKS

Centers for Disease Control and Prevention (CDC)
Health-related hoaxes and rumors.

www.cdc.gov/hoax_rumors.htm

United States State Department International Policy: Counterterrorism
Counterterrorism information site for the United States Department of State. Contains official statements, fact sheets, and speeches.

www.state.gov/www/global/terrorism/index.html

United States State Department "Patterns of Global Terrorism" Reports
The Hellenic Resources Institute offers the 1993, 1994 and 1995 "Patterns of Global Terrorism" reports on their site. These reports list the terrorist incidents during the year, as well as background information on the perpetrators.

www.hri.org/docs/USSD-Terror/

United States Department of State Travel Warnings: Consular Information Sheets
Official United States Department of State travel advisories for each country in the world. Also contains brief information on medical facilities, crime, infrastructure, legal issues, embassy locations, and other information useful for travelers.

http://travel.state.gov/travel_warnings.html

The Canadian Department of Foreign Affairs and International Trade's Travel Information and Advisory Reports
Travel information and advisories from the Canadian government.

www.voyage.gc.ca/destinations/ menue_e.htm

The British Foreign and Commonwealth Office Travel Advice Notices
Material intended to advise and inform British citizens travelling abroad. Key official information from USIA concerning counterterrorism. Very similar to the United States Department of State Counterterrorism site.

www.fco.gov.uk

FEMA Backgrounder and Fact Sheet on Terrorism
Basic information on terrorism from FEMA. The fact sheet lists what to do before, during, and after a terrorist attack.

www.fema.gov/hazards/terrorism/

Canadian Security Intelligence Service 1996 Public Report
Canada's counterterrorism program.

www.csis–scrs.gc.ca

GOVERNMENT LINKS

Central Intelligence Agency
www.odci.gov
www.cia.gov

United States Department of State
www.state.gov

United States Department of Justice
www.usdoj.gov

Federal Bureau of Investigation
www.fbi.gov

United States Marshals Service
www.usdoj.gov/marshals/

Immigration and Naturalization Service

www.usdoj.gov/ins/ins.html

United States Department of the Treasury, United States Secret Service, Bureau of Alcohol, Tobacco and Firearms, United States Customs Service, Office of Foreign Assets Control

www.treas.gov/index.html

United States Department of Transportation

www.dot.gov

Federal Aviation Administration

www.faa.gov

National Security Agency

www.nsa.gov

United States Postal Inspection Service

www.usps.gov/websites/depart/inspect/

Canadian Security Intelligence Service

www.csis–scrs.gc.ca

ACADEMIC AND INSTITUTIONS

Center for Democracy and Technology's Counterterrorism Issues
The CDT's page contains information and analysis relating to counterterrorism legislation.

www.cdt.org/policy/terrorism/

Electronic Privacy Information Center Counter-Terrorism Proposals
The Electronic Privacy Information Center is a resource for information on counterterrorism proposals and the restriction of free speech.

www.epic.org/privacy/terrorism/

Enough Terrorism
A project associated with the Terrorism Research Institute to educate the public and help stop terrorism and gang and drug violence.

www.terrorism.org

International Association of Counterterrorism and Security Professionals
The IACSP site has information concerning their organization as well as features from their magazine, *Counterterrorism and Security.*

www.worldonline.net

Censorship and Privacy: Terrorism Hysteria and Militia Fingerpointing Archive

The Electronic Frontier Foundation has information regarding legislative counter-terrorism efforts that use terrorism and militias as excuses for censorship.

www.eff.org

Anti-Defamation League

The ADL is the leading organization against anti-Semitism and its site has information on terrorism, hate crimes, skinheads, militias, and extremists.

www.adl.org

COUNTERTERRORISM LINKS

Terrorist Groups Profiles

Information about groups from the Naval Postgraduate School.

http://web.nps.navy.mil/~library/tgp/tgpndx.htm

The ERRI Counterterrorism Page

The Emergency Response and Research Institute has an excellent collection of news and summaries of worldwide terrorism events, groups, terrorist strategies, and tactics.

www.emergency.com/cntrterr.htm

Package Bomb Indicators

Information concerning mail-bomb indicators, courtesy of the ERRI and the United States Postal Inspector's Office.

www.emergency.com/pkgbomb.htm

Kim-spy: Paramilitary and Terrorism

The Kim-spy site contains a wide variety of links to paramilitary and terrorism information on the Internet.

www.kimsoft.com/kim-spy2.htm

Terror in Dhahran: Saudi Bomb Blast

CNN provides information on the blast, investigation, victims, and impact and analysis.

http://cnn.com/WORLD/1996/saudi.special/index.html

Olympic Park Bombing Special Section

CNN provides information on the blast, investigation, victims, and impact and analysis.

http://www.cnn.com/US/9607/27/olympic.bomb.main/index.html

The Police Officer's Internet Directory

A large collection of information relating to law enforcement.

www.officer.com

BIOLOGICAL TERRORISM LINKS

Centers for Disease Control and Prevention

www.CDC.gov

Emergency First Responder Equipment Guides, October 2001.

The National Institute of Justice (NIJ) is creating a series of guides about first-responder equipment that provide agencies with information on the types and capabilities of available equipment. While only the first guide in the series has been published, NIJ is releasing working drafts of the remaining guides in response to the September 11th terrorist attacks as soon as they are available. These draft guides are only available electronically on NIJ's website and are subject to change prior to being posted in their final form.

Selection of Chemical Agent and Toxic Industrial Material Detection Equipment for Emergency First Responders, NIJ Guide 100-00

An Introduction to Biological Agent Detection Equipment for Emergency First Responders, NIJ Guide 101-11 (draft)

Guide for the Selection of Chemical and Biological Decontamination Equipment for Emergency First Responders, NIJ Guide 103- 00 (draft)

Guide for the Selection of Communication Equipment for Emergency First Responders, NIJ Guide 104-00 (draft)

Guide for the Selection of Communication Equipment for Emergency First Responders
http://www.ojp.usdoj.gov/nij/pubs–sum/191160.htm

American Red Cross—Terrorism: Preparing for the Unexpected

www.redcross.org/services/disaster/keepsafe/unexpected.html

Federal Response Plan, April 1999

Includes revised Terrorism Incident Annex.

http://www.fema.gov/rrr/frp/frpintro.shtm

Texas Engineering Extension Service, Texas A & M University

http://teexweb.tamu.edu/

New Mexico Institute of Mining and Technology, Energetic Materials Research and Testing Center

http://www.emrtc.nmt.edu/

CalPoly Chemical and Biological Warfare Course
http://projects.sipri.se/cbw/

FEMA/National Fire Academy Course "Emergency Response to Terrorism: Incident Management"
www.usfa.fema.gov/nfa

FEMA/Emergency Management Institute Course "IS 195–Basic Incident Command System"
www.usfa.fema.gov/nfa

CRITICAL INFRASTRUCTURE PROTECTION LINKS

MIPT—Oklahoma City National Memorial Institute for the Prevention of Terrorism
www.mipt.org

United States Department of Defense
www.defenselink.mil

Office of Justice Programs, Office for State and Local Domestic Preparedness Support
www.ojp.usdoj.gov/osldps/

United States Army Soldiers and Biological Chemical Command
www.sbccom.apgea.army.mil/

State Health Departments
www.cdc.gov/search2.htm

Alaska Department of Health Biological Terrorism Information
http://www.chems.alaska.gov/bioterrorism_home.htm

Health Care Association of Hawaii, Emergency Management Program
www.hah-emergency.net

Los Angeles County Department of Health Services
(See biological terrorism subheading)
http://lapublichealth.org/acd/

New Mexico Weapons of Mass Destruction Preparedness
http://www.wmd-nm.org

New York City Department of Health
http://www.ci.nyc.ny.us/html/doh/home.html

Texas Department of Health Biological Terrorism Information
http://www.tdh.texas.gov/bioterrorism

Metropolitan Medical Response System, Field Operating Guide
www.ndms.dhhs.gov/CT_Program/Response_Planning/response_planning.html

ANTHRAX ARTICLES ON THE WEB

Anthrax. Dixon, T. C. et al., *New England Journal of Medicine*, September 9, 1999.
content.nejm.org/cgi/content/full/341/11/815

Anthrax as a biological weapon: Medical and pubic health management. Inglesby, T. V. et al., *JAMA,* May 12, 1999.
jama.ama-assn.org/issues/v281n18/ffull/ jst80027.html

Inhalational anthrax: Epidemiology, diagnosis, and management. Shafazand, S. et al., *Chest*, November, 1999.
www.chestjournal.org/cgi/content/full/ 116/5/1369

Bioterrorism alleging use of anthrax and interim guidelines for management. *MMWR*, February 5, 1999.
www.cdc.gov/mmwr/PDF/wk/ mm4804.pdf

ANTHRAX WEBSITES

ASAP at GW Medical Center: Special Lectures and Press Conferences
www.gwumc.edu/asap/index2.htm

AMA Message to Physicians on Anthrax
www.ama-assn.org/ama/pub/category/6383.html

CDC Health Advisory: How To Handle Anthrax and Other Biological Agent Threats, October 12, 2001
http://www.bt.cdc.gov/DocumentsApp/Anthrax/10122001Handle/ 10122001Handle.asp

Centers for Disease Control and Prevention: Anthrax
http://www.cdc.gov/ncidod/dbmd/diseaseinfo/anthrax_g.htm

SMALLPOX ARTICLES ON THE WEB

Smallpox: Clinical and epidemiologic features. Henderson, D. A., *Emerging Infectious Diseases*, July–August 1999.
http://www.cdc.gov/ncidod/EID/vol5no4/henderson.htm

Smallpox as a biological weapon: Medical and public health management. Henderson, D. A., et al., *JAMA*, June 9, 1999.
jama.ama-assn.org/issues/v281n22/ffull/jst90000.html

SMALLPOX WEBSITES

Vaccinia (smallpox) Vaccine Recommendations of the Advisory Committee on Immunization Practices (ACIP)

www.cdc.gov/mmwr/preview/mmwrhtml/ rr5010a1.htm

www.cdc.gov/mmwr/PDF/RR/ RR5010.pdf

CDC: Smallpox

www.bt.cdc.gov/Agent/Smallpox/Smallpox.asp

Medical Management of Biological Casualties Handbook, Chemical and *missing URL??*

PLAGUE ARTICLES

Plague as a biological weapon: Medical and public health management. Inglesby, T. V. et al., *JAMA*, 2000.
jama.ama-assn.org/issues/v283n17/ffull/jst90013.html

PLAGUE WEBSITES

CDC: Plague

www.bt.cdc.gov/Agent/Plague/Plague.asp

Basic Laboratory Protocols for the Presumptive Identification of Yersinia Pestis (Plague)

www.bt.cdc.gov/Agent/Plague/ Plague20010417.pdf

GENERAL INFORMATION

American Academy of Pediatrics. Chemical-biological terrorism and its impact on children: A subject review. *Pediatrics*, March 2000
www.aap.org/policy/re9959.html

American College of Physicians—American Society of Internal Medicine Bioterrorism Resources

www.acponline.org/bioterro/ index.html?hp

American Society of Microbiology: Resources Related to Biological Weapons Control and Bioterrorism Preparedness

www.asm.org/pcsrc/bioprep.htm

CDC MMWR: Recognition of Illness Associated with the Intentional Release of a Biologic Agent

www.cdc.gov/mmwr/preview/mmwrhtml/mm5041a2.htm

CDC: Agent List—Biological Diseases/Chemical Agents

www.bt.cdc.gov/Agent/AgentList.asp

CDC Health Advisory: How To Handle Anthrax and Other Biological Agent Threats, October 12, 2001.
www.bt.cdc.gov/DocumentsApp/Anthrax/10122001Handle/10122001Handle.asp

Medical Management of Biological Casualties Handbook, Chemical and Biological (Defense Information Analysis Center)

www.nbc-med.org/SiteContent/HomePage/WhatsNew/MedManual/Sep99/Current/sep99.htm

The Center for Civilian Bio-Defense Studies, John Hopkins University

www.hopkins-biodefense.org

CDC—Bioterrorism Preparedness and Response

http://www.bt.cdc.gov

The Center for Research on the Epidemiology of Disasters

www.md.ucl.ac.be/cred/front_uk.htm

The Defense Threat Reduction Agency—Chem-Bio Defense

www.dtra.mil

The Department of Health and Human Services, Office of Emergency Preparedness, National Disaster Medical System

ndms.dhhs.gov/NDMS/ndms.html

The Department of Health and Human Services, Office of Emergency Preparedness (OEP)

ndms.dhhs.gov

The Department of Defense, Nuclear, Biological, Chemical Medical References

www.nbc-med.org/others

Disaster Medicine and Medical Health

www.mentalhealth.org/cmhs/EmergencyServices

The Disaster Center

disastercenter.com

Center for Nonproliferation Studies
www.cns.miis.edu

International Association of Emergency Managers
www.iaem.com

International Critical Incident Stress Foundation (ICISF)
icisf.org

United States Army Medical Research Institute of Infectious Diseases (USAM-RIID)
www.usamriid.army.mil

United States Army National Guard Bureau
www.ngb.dtic.mil

United States Army Soldier and Biological Chemical Command (SBCCOM)
www.sbccom.apgea.army.mil

Anthrax (Department of Defense)
www.anthrax.osd.mil

STATE EMERGENCY OPERATION PLANS

Arizona
www.dem.state.az.us/serrp

Arkansas
www.adem.state.ar.us

California (Plans and Publications)
www.oes.ca.gov

Colorado
www.dlg.oem2.state.co.us/oem/Publications/BASPLAN.pdf

Delaware
www.state.de.us/dema/EMPLANS/deop/basic.pdf

Florida
www.dca.state.fl.us/bpr/Projects/ CEMP%20Online/cemp2000.htm

Hawaii (request via email)
scdwebmaster@scd.state.hi.us

Kansas (request via email)
Jleichem@AGTOP.STATE.KS.US

Kentucky
www.kyem.dma.state.ky.us

Louisiana
199.188.3.91/Plans/LAEmergAssistDistAct.pdf

Michigan
www.msp.state.mi.us/division/emd/publst.htm

Minnesota
http://www.dps.state.mn.us/emermgt/eop/index.html

Nebraska
www.nebema.org/opns.html

New Jersey
www.state.nj.us/lps/njsp/ems/ems.html

New York
http://www.nysemo.state.ny.us/planning.html

North Carolina
www.dem.dcc.state.nc.us

South Carolina
www.state.sc.us/epd

New Mexico
www.dps.nm.org/emergency/Plan/cover.htm

Texas
www.capitol.state.tx.us

Utah (request via email)
frontdsk@dps.state.ut.us

Virginia
www.vdem.state.va.us/library/eplan.cfm

Washington
www.wa.gov/wsem/2-ops/ops-plans/eop/eop-idx.htm

APPENDIX F

Orange County Fire Rescue
Emergency Communications Center
Standard Operating Procedure

The following evacuation plan is from the Orange County Fire Rescue Emergency Communications Center. In order to protect the confidentiality of this evacuation plan, specific phone numbers and locations have been ommitted. What this plan will show you is a basic format that your center can use to create your own plan.

SECTION 6.7 PSAP EVACUATION

EXECUTIVE SUMMARY

The attached plan provides the framework for the evacuation of the Orange County Fire Rescue Communications Center.

This plan assumes the following actions:

1. There is a five minute window between the emergency occurrence and the need to physically evacuate the building.

2. That we will use OFD as a call take center for our 911 calls, but we will use the Comm Mobile for all dispatch operations.

We still have to complete additional elements of the plan (see the last page for a list of things "to do").

We also have to write additional annexes to this plan for the following situations:

a. How the plan is changed when there is no five minute window of time for initial notifications.

b. How the plan is changed when there is a catastrophic event that wipes out this building and the on-duty personnel are not available to staff the Comm Mobile & OFD.

We will field test this plan during the Hurricane Zeek drill. Following that drill, we will make additional modifications to the plan and then set up training for all personnel.

GOALS OF A EVACUATION PLAN

This plan is designed to provide for the timely and safe evacuation of the police department communications center; and to designate accountability for training and testing of this plan on a regular basis. This policy shall designate the following activities:

a. Clear Authority to order an evacuation

b. Personnel safety issues

c. Field unit continuity

d. 911 Call Continuity

e. Notification to Command Level personnel

f. Pre-alert to the backup/alternate communications site

g. Regular Review, Training, and Testing of the Plan

HAZARD/THREAT ASSESSMENT

Any of the following emergency situations are possible in our area and could require the evacuation of the police department's communications center:

a. A bomb explosion in the public safety building

b. A fire starts on the floor below the communications center

c. Major damage occurs to the facility due to a tornado

d. A severe lightning storm or other event knocks out the facilities electrical system

e. A bomb or suspicious device is located in the building

f. A serious hazardous material spill occurs on the street in front of the facility

TYPES OF EVACUATION:

There are four types of evacuation that supervisors and assistant supervisors need to be familiar with:

1. Planned

2. Imminent

3. Immediate

4. Incomplete

Planned Evacuation

An anticipated event in which the public safety agency is in substantial control over the sequence and timing of the evacuation. Example: Hurricane

Imminent Evacuation

An urgent event where the communications center has a "window" (typically 5 minutes of less) to prepare and make notifications. Example: A fire in a section of your building.

Immediate Evacuation

A major event requiring the immediate evacuation of the Communications Center with no time to make notifications or preparations. Example: Tornado strike to your building, fire, explosion.

Incomplete Evacuation

A catastrophic event where the communications center facility is damaged or destroyed and the on-duty personnel are not able to relocate to another site to continue operations. Example: bomb explosion destroying your facility.

PRE-EVACUATION ISSUES

There are four required elements that must be in place to insure the success of the evacuation plan:

1. Planning

    Communications administration is responsible for the preparation of the evacuation plan and for coordinating the plan with the Sheriffs Office Communications Center.

2. Technology

    the Communications Division Technical Services Unit is responsible to insure that the necessary technology and system options are in place to provide for continuity of radio and phone systems.

3. Documentation

    Communications administration is responsible for documenting the evacuation plan and distributing it to all public safety entities who may need to use the plan; and for annual updates to the plan and the checklists.

4. Training/Drills

    The Communications Division Training Unit is responsible for providing annual training to all communications employees on the contents of the plan and to arrange a quarterly drill for each of the four communications squads.

EVACUATION PROCEDURE

Checklists for this procedure are attached to the policy and are used at time of evacuation.

The procedure below is written for an IMMINENT EVACUATION where there is a five minute window for notifications. Supervisors will modify the procedure to account for the need to immediately evacuate. PERSONNEL SAFETY IS FIRST--NOT NOTIFICATIONS!

Key steps for evacuation are as follows:

1. Supervisor on duty <u>assesses threat and makes decision</u> to evacuate, deciding Type of Evacuation and Rally Point.
(suggested Rally Point:

2. <u>Supervisor announces decision</u> to staff, and makes building-wide announcement on public address system.

3. <u>Time permitting</u>, some of the following functions may be performed: (personnel safety is first priority, not notifications!)

    a. **CALL TAKE 1** enters an emergency call for a full complement or higher level response as indicated by the type of emergency.

    CALL TAKE 1 notifies Orlando FD and asks them to <u>activate OCFRD evacuation plan</u> and reroute phones; and evacuates.

    If time permits, other phone lines are <u>manually transferred</u> or Bell South is notified to transfer other lines.

    b. **SUPERVISOR** uses speed dial button to notify Bell South to reroute 911 calls to OFD. **SUPERVISOR** notifies Command Staff and Comm Mobile Team personnel via group page using preformatted button; grabs <u>cellular phone</u> and <u>evacuation kit</u> and prepares to evacuate.

    c. **TAC-1 OPERATOR** will use Multi-Select 1 Function (M-SEL1) [which will select FIRE 1 through FIRE 15] and will activate the ALL STATION paging button [which will activate all stations] and dispatches the emergency event.

    d. At completion of the voice dispatch, the operator will announce the following:

    "Attention all stations, Orange County Fire Rescue evacuating the Comm Center, All units switch to and remain on FIRE 1, OFD will assist you"

    e. **NORTHSIDE AIP OPERATOR** generates CAD system printout for current unit activity, retrieves report from printer, grabs portable radio and evacuates.

    f. All other personnel not assigned a function evacuate.

    Personnel should take the portable radio at their position as they leave.

    g. In-progress emergency calls are transferred to OFD.
    Non-emergency telephone calls are terminated.
    Ringing phones are not answered.

4. Supervisor verifies that <u>all personnel did exit</u> the communications center and <u>secures facility</u> (security vs. emergency).

5. <u>Supervisor reports to the predesignated rally point:</u>

a. Verifies all employees are accounted for (notify District Chief of missing or injured employees)

b. Determines which actions were accomplished on the Evacuation Checklist.

c. Telephones OFD via cellular to request necessary follow-up action.

d. Arranges for travel to Alternate/Backup Site

6. <u>Supervisor will designate assignments</u>

a. Assign D2 and one additional employee to respond to OFD to perform liaison and call take functions.

b. Assign remaining employees to respond to the Comm Mobile at Station XX.

c. Travel is by employees personal vehicle.

7. <u>Arrive at OFD:</u>

a. Receive briefing from OFD supervisor

b. Verify what has been done on checklist

c. Determine which calls are in progress or holding for dispatch.

d. Assume TAC1 radio responsibilities until Comm Mobile is operational.

8. <u>Arrive at Comm Mobile</u>:

a. Activate and power up unit if this function was not already performed by the Station XX crew.

b. Call OFD and advise them of arrival at Comm Mobile.

c. Verify operation of all phones and radios using Comm Mobile procedure.

d. Notify OFD that Comm Mobile is ready to assume command of OCFRD dispatch operations.

e. Perform roll call of all emergency units to verify their status.

f. Handle dispatch operations per Comm Mobile procedures.

9. <u>On-duty personnel should notify their families that they are OK</u>
(Let employee call or have someone call on behalf of employee)

10. <u>Notify the oncoming Communications Center shift</u> of incident and change in reporting time and location

11. Notify <u>CISD as appropriate</u>

12. Consult with Command on <u>media issues</u>

SECTION 6.7 PSAP EVACUATION CHECKLISTS
SUPERVISOR FUNCTIONS

--->   Announce EVACUATION & RALLY POINT @

( )   Make announcement on **BUILDING WIDE INTERCOM**, press "INTERCOM" and "BLDG PAGE" button.

( )   Press "**ANSWER DID**" button and **SPEED DIAL** button labeled "_____" to broadcast evacuation signal

( )   Press "ANSWER DID" and then press SPEED DIAL button labeled "BELL SOUTH 911" and request that the operator activate ALTERNATE ROUTING OF 911 LINES TO ORLANDO FD PSAP.

( )   Grab **CELLULAR PHONE** from Supervisor Console

( )   Grab EVACUATION KIT located underneath Fire Alarm panel.

( )   Verify all employees did **EXIT** the Comm Center and assist as required.

( )   Close (secure) doors upon exit & **move to RALLY POINT**

( )   At RALLY POINT, verify that all **employees are accounted** for

( )   **Report any injured** or missing employees to the responding District Chief.

( )   Review what items were accomplished on **checklist** and **call OFD** to request follow-up on needed action.

( )   **Assign D2 and one additional** employee to travel to the OFD PSAP.

( )   **Assign remaining staff** to travel to Comm Mobile at Station XX.

CALL TAKE I OPERATOR

( )   Enter the emergency incident into CAD

( )   Press "**ANSWER DID**" and "**OFD**" to alert Orlando FD PSAP of the emergency: Tell OFD

This is an emergency !

Orange County Fire Rescue is evacuating their PSAP

Activate the OCFRD Evacuation Plan

TIME PERMITTING, THE FOLLOWING FUNCTIONS SHOULD BE PERFORMED

TRANSFER

DONE	LINE	CODE TO PHONE #	DESTINATION
( )	AFA LINE		Orlando FD Emergency Line
( )	Supervisor		Comm Mobile
( )	Business		Comm Mobile

SPRINT may be called to perform this function, at 1-800-xxx-xxxx.

TAC I OPERATOR

( ) Dispatch the emergency incident at OCFRD Headquarters

( ) Press Multi-Select function **M-SEL 1**

( ) Press **ALL STATION** page button

( ) Press **SEND**

( ) Dispatch assigned units to respond to OCFRD HQ

( ) Add following announcement at end of dispatch:

"Attention all stations, Orange County Fire Rescue evacuating the Comm Center, All units switch to and remain on FIRE 1, OFD will assist you, Units limit voice traffic to urgent transmissions until further notice."

( ) Grab portable radio from charger

( ) Evacuate

NORTHSIDE / AIP OPERATOR

( ) Printout CAD system activity

Go to _____ Menu

Use Function _____ labeled "_____"

( ) Go to computer room and retrieve printouts

( ) Close door to computer room when exiting

( ) Grab portable radio and EXIT

PSAP EVACUATION KIT CONTENTS

The Communications Center Supervisor assigned as the Comm Mobile Coordinator, will verify the contents of the Evacuation Kit each month.

ITEM	HQ	COMM	MOBILE	OFD
1. Evacuation Procedure and Checklists	X	X		X
2. OCFRD Employee Telephone list printout	X	X		X
3. OCFRD Building Master Key	X			
4. Keys to Comm Mobile	X			X
5. Backup file printout	X	X		X
6. OCFRD internal telephone directory	X	X		X
7. OCFRD Cellular Telephone Directory	X	X		X
8. OCFRD Paging Directory	X	X		X
9. Disaster Recall List of Orange County officials	X	X		X
10. Printout of ZONE file		X		X
11. Printout of GeoFile		X		X

PROCEDURE FOR ORLANDO FD SUPERVISOR

1. This checklist is designed to be activated upon request by the OCFRD Communications Center **OR** whenever the OFD Supervisor becomes aware of a major emergency occurring at the OCFRD Headquarters.

2. An OCFRD EVACUATION KIT will be maintained at the Orlando FD PSAP.

( ) Receive request to activate plan

( ) **Verify that HELP has been dispatched** (ie: Fire/Rescue & OCSO)

( ) Call **BellSouth 1-800-xxx-xxxx** and request **ALTERNATE ROUTING** of Orange County Fire PSAP 911 lines to Orlando FD

( ) Notify **OCFRD** personnel using **ALPHA PAGE**

( ) Assign operator to **monitor OCFRD TAC 1** to process urgent requests for assistance, until OCFRD personnel arrive at OFD.

( ) Anticipate phone call from OCFRD Supervisor via cell phone shortly after evacuation

( ) Anticipate arrival of two OCFRD personnel at OFD to handle OCFRD call taking and dispatch functions.

OFD LIAISON TEAM CHECK LIST

( ) Arrive at Orlando FD PSAP and **receive briefing** from on-duty Supervisor

( ) **Review OFD Checklist** to determine what action steps need to be completed

( ) Determine what **calls are in progress** or awaiting dispatch. Create list.

( ) Assign one operator to handle **TAC 1** radio traffic and dispatch emergency calls until Comm Mobile can assume this function.

( ) Assign one operator to handle **incoming calls** for OCFRD.

OCFRD CAD terminal may be used to geo-verify addresses and determine station order assignments. A PC file and hard copy file will also be provided in the event the CAD terminal is not functional.

Incoming calls for service are recorded on a manual card, along with station order and then phoned over to Comm Mobile for dispatch.

( ) Both employees should **call home** ASAP to advise they are OK

( ) Await notification that remaining personnel have arrived at the Comm Mobile and are preparing to assume dispatch responsibilities.

COMM MOBILE CHECKLIST

( ) Arrive at Comm Mobile

( ) Activate and Power Up unit if not already activated by Station XX crew [see Comm Mobile Procedure]

( ) Call OFD at xxx-xxxx and advise them at OCFRD team has arrived at the Comm Mobile and will begin checking equipment.

( ) Review OFD checklist to determine what has been done. Assign responsibilities for remaining functions.

( ) Verify operation of all phones and radios (use Equipment Checklist)

( ) Call OFD at xxx-xxxx and receive final update on unit status and calls holding

( ) Advise OFD that you are ready to assume responsibility command for OCFRD dispatch.

( ) Verify status and location of any unit/station not accounted for

( ) Follow Comm Mobile dispatch procedures (see Comm Mobile Dispatch Procedure)

( ) Have employees notify families that they are OK

( ) Determine if any employees need to be immediately relieved and call in additional employees to provide coverage.

( ) Notify oncoming communications shift supervisor of incident and determine where oncoming employees should report (ie: which employees to OFD and which employees to Comm Mobile).

( ) Notify CISD as appropriate.

( ) Consult with PIO on Media Issues.

APPENDIX G

Overview of Community Actions

The following is an overview of communication actions during the Murrah Federal Building Bombing. It covers what happened and recommendations for future incidents. This overview is from the Oklahoma Police Department Report.

COMMUNICATIONS UNIT (9-1-1)

The Emergency Management Unit is responsible for answering all 9-1-1 calls directed by Bell Telephone. Calls for Fire and Emergency Medical Services (EMSA) are immediately transferred to the respective dispatch unit. All calls for police service are processed and dispatched by unit personnel. All calls coming into the unit are recorded and documented utilizing a Computer Aided Dispatch (CAD) System.

Beginning April 19th, the 9-1-1 Center was inundated with calls. During the first hour of the incident, 488 calls were placed to police communications. Normal call volume activity on a Wednesday morning usually averages around 80 calls per hour. Normal 24 hour activity is around 1,800 calls. The 24 hour activity for April 19th was 2,969 calls. For the next three days, call volume consistently remained about 20 percent above normal. The call volume statistics are for 9-1-1 Center emergency lines only and do not include calls to administrative lines or emergency calls placed directly to fire or EMSA dispatch centers.

It should be noted the call statistics for the initial hour of the incident are deceptive in that the numbers reflect only the calls which were answered. During the first hour of the incident there were over 1,800 calls attempted on 9-1-1 lines alone. At least 1,212 callers received a busy signal due to all incoming trunk lines (15 9-1-1 lines, seven non-emergency lines and two overflow lines) all being used. This can not be avoided in future incidents without adding additional trunk line and additional personnel to man the lines.

Call volume jumped from 35 in the half-hour preceding the incident to 338 in the first half-hour of the incident. Of the initial calls, 83 9-1-1 calls were initially abandoned (caller hung up before a dispatcher could answer). Per existing unit policy, call backs to abandoned calls were made as off-duty personnel responded for duty. Two dispatchers were assigned to returning the abandoned calls with first priority given to those numbers in the downtown area as they were more likely to have been from bombing victims. Other abandoned calls were called back in a priority based on telephone prefixes as they radiated outward from the blast site. This process took several hours to complete and no bombing victims were located by these measures.

Other measures to handle the overload included immediate referral of all press and media calls to the PIO with a prompt disconnection. Routine calls for police service were reduced by immediately suspending the processing of all lesser calls, these being the priority 4, 5 and 6 calls. This measure was continued for approximately the first 24 hours of the incident.

Due to problems in making outgoing cellular telephone calls from the scene, it was necessary for Communications to keep one dispatcher assigned to monitoring an open-line to the Police Command Post. This assignment continued for several days due to the limited number of telephone lines and the volume of traffic to the Command Post. Many requests unique to this incident were made of dispatchers such as the need for dump trucks, portable restrooms, lighting equipment, heavy equipment, etc. These requests were coordinated with Fire Dispatchers and could have been expedited with an Emergency Resources Catalog listing contacts. Similarly, the Major Disaster Call-Out List had not been updated because the Emergency Management Coordinator position is vacant.

Since a crime was involved, criminal intelligence information began coming into dispatchers. Initially this was a concern since there was no centralized investigative command post. As the FBI set up operations, dispatchers referred information to Federal officials.

Other initial call demands included the activation of the Emergency Response Team (ERT). The activation required a dispatcher to set off the various team pagers prompting 70 officers to call into an already overloaded system. All ERT members had to call Communications to find out where they would stage. Additional confusion was created when a change in the staging location occurred.

On April 19th, 15 Dispatch personnel were on duty at the time of the incident. Within the first 90 minutes, 17 additional Dispatchers were either called in or voluntarily reported for duty. Eight others either reported early for Shift 2, stayed late or both. Had the communications crisis lasted for another 12 hours, it would have required personnel to go to 12-hour shifts in order to provide for rotation of personnel.

At the time of the April 19th bombing, Communications Center security was an issue. From the beginning of the incident, the electronic front gate was locked open since a security camera has a full view of the gate area. An offer for military police at the site was declined. Security for the center would have become a concern had there been a threat or had the initial crisis lasted longer than 12 hours. All tours and unauthorized personnel were denied entrance to the facility through May 5th.

Included in the 9-1-1 Center is the City Emergency Operations Center, (EOC). EOC routinely dispatches for other city departments, monitors weather conditions, coordinates needs for emergency crews with utilities and answers City Hall telephones after routine business hours. On April 19th, one employee was on duty in EOC.

Immediately following the blast, all dispatching for other City departments was suspended. Only emergency radio traffic was handled in an effort to coordinate needs at the scene. One of the more immediate problems encountered was when most of City Hall was closed. City offices were closed after the report of a second explosive device. All offices were evacuated and employees were directed to go home. The telephones at City Hall were forwarded to EOC, causing two immediate problems; difficulty in reaching City workers in some areas and the massive increase in call volume to EOC. Most of the calls were from citizens making inquiries or requests which could not be answered from EOC.

Increased calls to EOC added to the workload for dispatchers when two City hall telephone numbers were publicized for inquiries about persons who were in the Federal Building at the time of the bombing. Dispatchers were not notified in advance and had no instructions or information for the callers. This situation took nearly two hours to correct.

COMMUNICATIONS RECOMMENDATIONS

There were a variety of problems dealing with communications. Issues included the volume of calls to 9-1-1, radio traffic and coordination of assignments for responding units, other city departments, assisting agencies, and communicating with personnel on-site.

    a. The volume of calls at 9-1-1 presented problems in two areas. First, the volume of calls about the incident and officers activated created an overloaded system. Many callers reached busy signals, thus slowing the response for some officers. Secondly, the evacuation of City offices and forwarding of their telephones to EOC created an overload in this area.

It is recommended procedures for activation of officers be reviewed, see recommendations listed under personnel/assignments. An alternate site for forwarding of City office telephones needs to be identified or provisions made for City employees to aid in manning the lines in emergency situation.

    b. From the very beginning, responding police and fire personnel could not talk to Federal Agencies or one another. It was late on the first afternoon before a communications network could be established. In the interim runners were dispatched to the various command posts and to sites to deliver directives.

Recommend the Department provide portable data terminals and operators to all responding agencies command centers. All authorized actions could be broadcast on these data terminals to command centers for dissemination to their personnel. This would provide a secure network for communication between command centers. A radio channel could be identified for all City Department vehicles. This would permit communication with Traffic Control and Public Works. The acquisition of an 800 MHz trunking system for City Departments would provide access to a greater number of frequencies, permit channels to be dedicated to specific functions required for the incident and provide inter-departmental communication.

> c. Communication with personnel on-site was hampered by the lack of each officer having a hand-held radio. Some officers had pagers and/or cellular telephones.

It is recommended all officers working at the site report to the Command Post. This would ensure a means of communicating with the officers was established prior to the officer beginning an assignment.

The following is the Communications section taken from the Alfred P. Murrah Federal Building Bombing Final Report.[1]

OKLAHOMA CITY BOMBING FINAL REPORT: COMMUNICATIONS

The rescue and recovery effort utilized every available communications system. Landline, cellular, two-way radio, and digital communications were all vital in the first hectic hours of the disaster and throughout the days that followed. Each system provider maintained service, directed system resources to the rescue work, and provided countless hours of support and equipment.

The public and corporate communications systems that supported the effort include the City of Oklahoma City's two-way radio system, E-911 Communications, Computer Aided Dispatch, Southwestern Bell Telephone, AT&T Wireless Services, and Southwestern Bell Mobile Systems. Other metropolitan areas cities, Oklahoma County, and the various State agencies that supported the rescue work utilized their communications systems. The Urban Search and Rescue Teams brought in their own systems. FEMA brought in a Mobile Emergency Response System (MERS) that linked federal agencies sent to Oklahoma City with Washington, D.C.

---

1. Reprinted with permission from Fire Protection Publications/IFSTA and the Oklahoma City Fire Department.

CITY OF OKLAHOMA CITY COMMUNICATIONS

TWO- WAY RADIO SYSTEM

The Oklahoma City Fire Department maintains the City's two-way radio system. This system serves all departments which have mobile personnel and equipment, except for the MassTrans Bus System which has a private maintenance contract. It is the basic system for communications between headquarters' units and field personnel, and it allows users to hear all communications over their assigned channel. The system serves the Fire Department; Police Department; Emergency Operations Center (EOC); Public Works Department including Streets, Traffic Operations, Engineering, Inspection, and Sanitation Divisions; Department of Airports; Parks and Recreation Department; Building Management Division of General Services Department; Animal Welfare; Neighborhood Enhancement Services; Water and Wastewater Utilities Department including Water and Sewer Line Maintenance and Utility Customer Services Divisions; the MassTrans Bus System; and the Oklahoma City Zoo.

The City's two-way radio system is actually 35 systems, with the older systems operating in 150 Megahertz (MHZ) ranges and the newer ones in the 450 MHZ bands. The system utilizes repeaters, towers, and transmitters located across the city and linked by telephone lines. Radio transmissions go to three to six receivers and are transmitted to a voter by telephone line. The voter selects the best quality signal and sends it by telephone line to a transmitter site where it is transmitted over the air.

The system depends upon the integrity of the repeater and transmitter sites and of the telephone landlines. If the telephone/and lines or any of the major system components fail, direct communications from radio to radio is possible. In this circumstance, handheld transmissions would be limited to one-half mile and car radio transmissions to about 15 miles.

Each department's or major division's radios operate on a designated channel or channels. The channel selection for each radio is preprogrammed by Fire Communications. Traffic on each radio is limited to the designated channel or channels. This allows each department to control access to its radio communications but precludes direct radio communications between the various departments. The Emergency Operations Center, which is part of the Police Emergency Management Division, monitors traffic on the City's radio system. Units from on department can radio the EOC and have messages relayed to a unit in another department.

The two-way radio system provided dependable, but unsecured, voice communications especially in the early stages of the event when landline and cellular phones were overwhelmed with calls. Each department or division exercised control, limiting traffic to essential communications.

FIRE COMMUNICATIONS MAINTENANCE CENTER

The Fire Communication Maintenance Center is responsible for maintaining the City's two-way radio system. It is located at 600 N. Portland in the Fire Maintenance Center complex, which is adjacent to the Fire Training Center. The Chief Fire Communications Officer is Ron Rainbow.

The bomb blast was heard and felt at the Maintenance Center. Personnel went outside and looked to the east. They saw the column of smoke towering over downtown. Reports from those arriving at the blast site were heard over Fire's radio system. Personnel at Communications and Maintenance responded to the disaster. Chief Ron Rainbow dispatched his personnel to inspect each tower and transmitter site. None of the sites or the landline connections serving them were damaged. Throughout the incident, he had men and equipment on standby to repair any parts of the system.

One of Chief Rainbow's first concerns was to ensure there would be an adequate supply of handheld radios and batteries. The Department uses several different models of radios, each requiring a different battery. He immediately contacted the major suppliers to tell them what had happened and arrange for delivery of radios, batteries, and battery chargers. The vendors supplied about 900 radios and over 3,000 batteries.

A Communications Center computer program was used to program the new radios to designated channels. Fire Communications then delivered the radios and batteries to the Logistics Command Center, which was set up on 7th Street north of the Murrah Building.

Once Chief Rainbow had seen to the security and needs of the communications system, he assisted Chief Soos of Fire Maintenance in the general mobilization of supplies and equipment to support the rescue effort. This work continued until about 10:00 p.m. on April 19.

The two-way radio system experienced only one problem during the incident. For a brief period, Fire and Police were getting "cross talk," or hearing each other's communications on radios at the Murrah site. The cross communications came from the emergency telephone lines at the site. Southwestern Bell Telephone and Fire Communications quickly corrected this problem. Fire Communications did not have personnel at the Logistics Command Center where the handheld radio batteries and chargers were supplied. Many batteries were probably discarded because they were not properly charged before reuse.

When the Multi-Agency Coordination Center (MACC) was established at the Myriad Convention Center, Chief Rainbow was assigned to be the communications coordinator for all of the agencies. He was assisted by Major Larry Finch of his staff. They coordinated the provision of handheld radios and cellular phones.

The work of Fire Communications assured that adequate radios and batteries were available to support the rescue and recovery work and that the City's two-way radio system functioned reliably throughout the incident.

OTHER CITY RADIO

Each City Department used its radio channels to dispatch equipment and personnel and to maintain communications in support of the rescue effort. Public Works Chief Dispatcher Ron Stephens was assigned to the command post and later to the MACC operation at the Myriad Convention Center. He utilized the radio to order supplies and equipment for logistical support.

FIRE DISPATCH

Fire Dispatch is located in Station 1 at 820 NW 5ht Street. Its mission is to dispatch emergency vehicles on every call. The Dispatch Center has three consoles and each is equipped with a two-way radio, a telephone, and a console for the Department's Computer Aided Dispatch system. In a corner behind the consoles is a light board with a map of the city that shows all but the two newest fire stations.

When a station is responding to an alarm, a red light is displayed for that station. The Dispatch Center is staffed by five-person crews on 24-hour shifts. The Chief Dispatcher is Harvey Weathers.

The 9-1-1 Emergency System immediately transfers all fire calls to Fire Dispatch. The calls are answered by two dispatchers, one who talks with the caller and a second who listens to the call and enters the information into the Computer Aided Dispatch system that identifies the location of the incident, the map number of the incident, and the closest responding unit. Units are dispatched by audio and/or radio. They are given the map number to facilitate a quick response to each incident.

In addition to the direct lines from 9-1-1, the Dispatch Center answers two "Information Only" numbers published in the telephone directory and the seven digit Fire Emergency phone number that was used before the 9-1-1 system was put into operation in 1989. Fire's two-way radio system is assigned 12 channels and usually operates on channels one, two, and three. Channel selection is made by using a switch on the radio. Channel four is a mutual-aid channel that can be used to provide radio communications with other metropolitan area fire departments.

On the morning of the incident, Chief Weathers and five dispatchers were on duty. The morning had started as usual with a 7:00 a.m. audio test for all stations followed by a radio test for all the rigs. The bomb blast shook Fire Station No. 1, which is just six blocks west of the Murrah Building. None of the communications equipment was damaged. The command personnel and all truck and engine companies from Fire Station No. 1 self-dispatched toward the column of smoke from downtown.

The Fire Dispatch Center was active within seconds of the bombing. All the phones began to ring. The first calls were from security alarm companies. The force of the blast activated alarm systems that in turn rang Fire Dispatch. Alarms were activated in buildings as far as 16 miles from downtown. Other calls came from people wanting information about the noise they had heard.

Two-way radio was critical to fire communications in the early hours. Fire Dispatch's tape of April 19 records immediate radio traffic beginning with Engine 17 reporting a large column of smoke "south of this address." The District Chief reports he is establishing command at 6th and Harvey. The Research and Development Officer reports the need for multiple alarms and identifies the Federal Building as one of the damaged buildings. The Shift Commander orders Dispatch to start multiple Emergency Medical Services Authority (EMSA) units. The Fire Chief orders a stop to radio traffic and orders commands set up at various buildings damaged by the blast. He calls for a pumper to put out the car fires. The Fire Chief then orders a General Alarm for every available piece of equipment.

The radio channel in use at the time of the bombing was dedicated to the incident. Dispatch assigned one console to cover the incident channel. The dispatchers made sure that every request and every transmission to other companies was received. All non-incident traffic was put on another channel and monitored at a second Dispatch console.

The Fire Department's Incident Command System operated from a mobile Command Post at the Murrah site. The Command Post handled most of the communications at the site. Fire Dispatch sent one dispatcher to assist at the command post. Dispatch monitored the Command Post traffic. They noted the materials and equipment that were being requested at the site. Often they had identified a source and were ready to act when the directive came from the Command Post.

Throughout the incident, Dispatch's time was divided between support for the rescue and recovery work and receiving and dispatching non-incident calls. Dispatch received offers of help from metropolitan area fire departments and coordinated the placement of those units at vacant Oklahoma City stations to provide non-incident coverage for the 621-square-mile city. Maintaining control of what units were available at over 30 fires stations was a demanding duty for Dispatch. Two-way radio communication between Dispatch and the other fire departments was maintained by using channel four. The other fire departments do not have OCFD's detailed maps. When these departments responded to non-incident calls within Oklahoma City, Dispatch would talk them to the site over channel four.

By 10:00 a.m. on April 19, four off-duty dispatchers had reported for duty. All telephones lines rang constantly throughout the first days with requests for information and offers of assistance. Chief weathers directed the work of the Dispatch Center and assumed responsibility for coordinating offers of assistance. Dispatch maintained a log of offers of materials , supplies, and services and faxed the information to the command post. Over 40 double-sided pages of offers were recorded. He also took calls from the media who could not contact the Public Information Officer and provided information to them.

During normal operations, the Fire Department is the first responder to all calls for EMSA. Because of its more compact service areas, the Fire Department is usually able to reach the scene before the responding ambulance. On the day of or the day after the blast, EMSA contacted Fire Dispatch and recommended that the Fire Department not respond to these calls since so much of its equipment was at the Murrah site. The Fire Department accepted this recommendation and did not provide medical first responder service for about 22 hours.

During the incident, communications within the Fire Department were enhanced through the use of cellular phones. These phones were used for requests for manpower and supplies so that radio communications could be used either for crucial information or for information which needed to be heard by all users.

POLICE COMMUNICATIONS AND DISPATCH

The Police Communications Center is a part of the Department's Emergency Management Division under the command of Captain Ron Owens. The Communications Center is located at 4600 N. Martin Luther King Blvd. The Communications Center includes the City's E-9-1-1 and Police Dispatch Center and the Emergency Operations Center. The Center has 15 incoming 9-1-1 lines, seven non-emergency lines and two overflow lines.

The 9-1-1 Center answers all calls to the 9-1-1 Emergency number. Calls for the Fire Department and for EMSA are transferred to their respective dispatch centers. There is a direct ring-down line to the Fire Dispatch but not to EMSA. All calls for Police service are processed at the 9-1-1 Center using a Computer Aided Dispatch (CAD) system. The Dispatch Center monitors the police two-way radio system and has a dispatcher assigned to each of four patrol frequencies and one to the administrative frequency. The Center can communicate with the Fire Department by two-way radio and by digital communications through the computer aided dispatch system.

The Emergency Operations Center (EOC) monitors the City's other two-way radio channels and provides assistance as needed. The EOC is the dispatch center for the Animal Welfare Division. When the City's downtown offices were closed, the phones for the City Manager, Council Support Staff, and major departments and divisions are set to automatically ring at the EOC.

The 9-1-1 Center has 15 telephone consoles, each of which can answer 2 calls at one time. There are six two-way radios consoles. At the time of the bombing Lt Scott McCall was on duty as Supervisor. Allan Garlitz and Tony Harrison were the Communications Supervisors. In addition to the supervisors, twelve 9-1-1 dispatchers, two EOC dispatchers, and four staff people were on duty. Within the next 90 minutes, 17 additional dispatchers were called in or voluntarily reported for duty.

The first call reporting the bombing was received at 9:02:20 a.m. Within the next hour, there were over 1,800 call attempts on the 9-1-1 lines. Of these, 1,212 received busy signals because all of the lines were in use. The Center handled 338 9-1-1 calls within the first half hour and a total of 488 calls within the first hour. Normal activity for a Wednesday morning is 80 calls per hour. Most of the incoming calls were related to the disaster, but others were for unrelated incidents.

For the next three days, call volume was at 20 percent above normal. In addition to the 9-1-1 calls, the administrative lines were flooded with calls for information and offers of assistance.

The first eight to ten minutes were very frustrating as E-9-1-1 and EOC personnel did not know exactly what had happened. The noise and force of the Murrah explosion had been so great that citizen calls and police radio were reporting an explosion at points all across the city. One officer reported a bomb north of Lake Hefner, which is more than ten miles north of downtown. As Police cars approached the Murrah building and reported damaged buildings and walking wounded, it was clear the explosion was downtown, but the site was still not identified. Friends and relatives of downtown workers were calling 9-1-1 and the EOC for information, and the staff could not tell them what had happened or what buildings were involved. The explosion set off residential alarms all over the city. Once staff realized the magnitude of the blast, they canceled out the alarm calls because they did not have the manpower to check them out while handling the emergency.

While local media knew to call the City's Public Information Office, out-of-town media were calling the EOC and 9-1-1. The reporters were wanting to tape statements and demanding to stay on the line until they could get information. Staff needed to direct all resources to the disaster. They quickly adopted a policy of telling the media to call Public Information and then hanging up on those calls. Staff also began to receive calls from across the nation with offers of every kind of assistance.

During the first hours, 9-1-1 and the EOC handled a number of duties. Two dispatchers worked as a team to handle all the two-way radio traffic for the Police Will Rogers Patrol Division whose district included the incident area. They relayed vital information to and from the police officers at the site. Two other dispatchers were assigned all Will Rogers Patrol Division and Hefner Patrol Division traffic not related to the bombing. They assigned whatever units were available to handle those calls. Many off-duty police officers with take home cars called the Communications Center to place themselves on duty. Communications assigned many of the non-incident calls to them.

Two dispatchers were assigned to return 83 abandoned 9-1-1 calls. These were calls that were recorded on the system but which hung up before they could be answered. There was concern that calls from the downtown area could have been from victims needing assistance. All abandoned calls were called back, starting with those from downtown. This process took several hours, and it did not identify any persons requiring help.

The Center was required to activate the Police Emergency Response Team (ERT). Phone calls had to be made to team beepers programmed to activate over 70 team members. Each team member then called Communications to find out where to report. There was a problem making these pages as the phone lines were always busy. The ERT members in turn had difficulty making calls to the Center to receive information on where to report for staging.

Until landline phones could be installed, the Police Command Post used cellular phones. The cellular system was overloaded. Because the Command Post could not be sure of getting a line out, they never broke the first connection with 9-1-1. For the first 24 hours, the Center placed on dispatcher in charge of monitoring the open line. Requests for materials, equipment, and other support were sent to 9-1-1 over the open line and then dispatched to City departments and other agencies that could respond. Some communication with Command Post was established using a mobile data terminal.

The EOC dispatchers quickly became involved in locating equipment and supplies that were needed at the incident and in dispatching requests to City departments. They were instrumental in locating barricades, fencing, and dozens of other materials needed at the site. They also began to take the calls offering assistance and developing a log of items that might be needed. The City's downtown offices were closed at about 10:30 a.m. because of the threat of a second bomb. EOC staff handled all of the calls that automatically rang to the EOC.

The Oklahoma State Bureau of Investigations (OSBI) was working to compile a list of all of the people who had gotten out of the Murrah Building. At the press conference held on the evening of April 19, all people who had gotten out of the building or who knew someone who had were asked to call City Hall. Because if was after hours, these calls came to the EOC even though people were on duty downtown. EOC personnel took on the duty of answering these calls and compiling a list of those known to be out of the building. They compared this to the list of persons who worked in the Murrah Building and crossed of the names of those who had gotten out. This helped create the list of persons still needing to be accounted for.

The 9-1-1 Center had a direct ring-down circuit to the Fire Department but not to EMSA. Because so many citizens were calling all of the emergency services, E-9-1-1 dispatchers experienced difficulties in transferring calls to the EMSA Dispatch Center. This problem continued for several hours after the bombing, high lighting the need for a computer or radio link between them.

By the time the shift changed at 3:00 p.m. on April 19, they had received over 1,500 enhanced 9-1-1 calls. The 9-1-1 call total for the first 24 hours was 2,969 calls with an additional 700 calls to the EOC. After the first 24 hours, most Police communications for the incident went through the Command Post, lessening the impact on the Communications Center.

For the first 24 hours, routine processing of Police priority 4, 5 and 6 calls was discontinued so that Communications staff could deal with disaster related calls. The EOC did not respond to the Animal Welfare calls. The Communications Center continued to receive a high volume of calls throughout the incident. When the site became a crime scene, citizens called with investigation information. Once the FBI was involved, these calls were directed there. The Communications Center was the focal point of all kinds of calls. The Center did not have direct communication links with other agencies and in many cases did not have direct phone numbers for many officials. A great deal of time was spent trying to get information or requests to the right person. The relocation of the MACC to the EOC was a great asset to the Communications Center as calls or information could be given to an agency representative who would handle it from there.

The Communications Center did an excellent job of handling 9-1-1 calls and supporting the Police Department and the rescue and recover effort. Communications staff were subject to a great deal of stress both from the long hours of work and from being closely related to the disaster without being able to directly participate in the rescue work. Most staff took advantage of stress-relief sessions offered to all those associated with the incident.

MOBILE DATA TERMINALS
Over 800 Mobile Data Terminals (MDT's) are in use in Police and Fire vehicles. These provided secure messaging capability during the rescue effort and were critical to the exchange of sensitive information. The Police Mobile Command Post had six live MDT's. These were used to communicate with vehicle-mounted MDT's and some handheld MDT's throughout the rescue effort.

CITY COMMUNICATIONS
The Management Information System (MIS) Division of Oklahoma City's Finance Department is in charge of landline phone and computer networked communications within city government. Technical Support Manager Lucien Jones was in

charge of the Communications Section. The City uses a centrex phone system that allows four-digit calling on over 2,500 lines within the system.

The switch for the system is located in one of the Southwestern Bell downtown buildings. The City has an extensive fiber-optic computer network that encompasses all of the downtown office buildings including Police Headquarters and Fire Station1.

Jones is a member of the Oklahoma Disaster Preparedness Council and had participated in emergency communications planning for Oklahoma Cit. Jones is a former police officer and closely involved in all areas for Police communications. He knew his role would be to maintain the City's internal communications systems and to coordinate City needs with the corporate communication providers including Southwestern Bell Telephone and AT&T Wireless Service.

Jones was at his office in the 100 N. Walker Building when the blast occurred. The explosion shook the seven-story building, which was undergoing extensive renovation. Jones and the others in the building rushed outside, thinking that there had been a construction accident. Once Jones saw the smoke from the explosion, he returned to his office to check on the communications system. The computer and phone networks were not damaged. They functioned without problems during the incident.

Sometime after 10:00, Jones and MIS Director Kerry Wagnon proceeded on foot to the blast area. When they neared the site, they were met by the people rushing to evacuate the area because of the threat of a second bomb. Jones eventually worked his way to the interim Police Command Post site at NW 8th and Harvey. Jones learned that the Police Mobile command post urgently needed new landline phone umbers. The mobile command post had a "cellular" network, which allowed the phones to operate over the cellular network. The media, which used scanners to listen to Police radio communications, had learned the phone number of the Command Post and broadcast it to citizens as a number to call for information about missing or injured persons. The Command Post did not have this information. The flood of citizen calls disrupted Command Post communications.

Jones had a long established working relationship with Southwestern Bell. He requested that the first available lines be connected to the Command Post. Bell ran the closest four lines into the Command Post. All of the numbers were active numbers; two of them were to nearby doctor's office and one was to a church parsonage. The numbers rang at both sites, prompting some confusion. But they enabled the police to disregard the number broadcast to the public and to have landline communication in and out of the Command Post. When the police Command Post was relocated to the Southwestern Bell Headquarters site, new phone lines and numbers were provided. As a result of this problem, the Police Department never again broadcast the Command Post phone numbers. Instead, the numbers were released through computer aided dispatch and were given out at Police shift briefings.

Jones provided continuing assistance to the Police Command Post. He coordinated phone placement orders with Southwestern Bell Telephone. He coordinated efforts between AT&T Wireless Services and the Police Department to locate a Cell on Wheels site N. Walker. He was in charge of relocating the MACC phone system from the Myriad Convention Center to the EOC.

CORPORATE COMMUNICATION SYSTEM

The corporate communication provider in Oklahoma City gave every possible support to the rescue effort. Their assistance began within minutes of the bombing and continued throughout the incident. Landline cellular and long distance services were provided to every department agency and non profit group involved in the rescue work or in serving the victims and their families.

SOUTHWESTERN BELL TELEPHONE

Southwestern Bell Telephone Company (SWBT) is the franchised landline telephone provider for Oklahoma City. Its corporate headquarters, known as One Bell Central, are located in the former Central High School building between NW 7th and NW 8th Streets, just two blocks north of the Murrah building. This four story building fronts on Robinson Avenue. On the rear or Harvey Avenue side, there is a covered parking area adjoining the building and a large surface parking lot. Additional surface parking lots are located across Harvey Avenue. SWBT has personnel and equipment in several other downtown buildings, including the building just south of One Bell Central.

The bomb blast caused considerable damage to all of SWBT's downtown buildings. The glass was blown from most of the windows on the north and west sides of One Bell Central, and there was other damage to the building. The damage and the threat of another bomb caused SWBT to evacuate One Bell Central and its other buildings. The voice mail system for the vacated offices continue to function. This allowed relocated SWBT staff to monitor and return calls from agencies needing their assistance.

With two hours of the blast, more than 12 million calls were attempted in Oklahoma City. This is three times the normal volume. Because so many people were attempting to use their telephones, some callers did not get a dial tone on the first attempt. SWBT immediately implemented emergency network controls. Disruptions in service were minimal and were limited to the early hours of the disaster. By mid-afternoon on April 19, there was little, if any, delays in call completion.

Southwestern Bell Company President David Lopez offered all of his company's resources to the rescue and recovery effort. SWBT contributions would include: use of One Bell Central's covered and surfaced parking areas as the site for Command Post vehicles; provision of telephone lines to all Command Post vehicles; use of One Bell Central as a dormitory area for two USAR teams and other rescue support functions;

provision of telephone lines into the various command and logistic sites within the perimeter including the Murrah loading dock area; provision of telephones and switching equipment for the MACC at the Myriad; provision of telephone lines to the media area; computer aided design support for the rescue and recover work; protection of the climate-sensitive E-9-1-1 call processing computers, which were located in one of SWBT's damaged building; and mobilization of volunteers who assisted in all areas. Southwestern Bell Mobile Systems, a division of Southwestern Bell Communications Corporation, provided cellular communications support. This work is described in the Cellular Communications portion of this section.

Telephone lines for the Command Post vehicles parked at One Bell Central were provided by extending phone lines from the building. In many cases, a working phone line was identified in an office and ordinary telephone wire was extended from the phone jack, out the office, down the corridors, out the back door, and into the Command Post. These lines were subject to a great deal of wear and tear and required frequent maintenance. Because the lines were run at random, there were no hunt groups for agencies with several phones. Some phones, if not answered in several rings, would go to the former user's voice mail. These were minor problems that were more than compensated for by Bell's prompt provision of these phone lines.

The above and underground telephone lines serving the buildings in the immediate bombing site were severely damaged or destroyed. Because all of the building in the are were evacuated, emergency restoration of service to those buildings was not required. However, secure telephone service was needed for all the agencies participating in the rescue and recovery work within and around the perimeter and for the media area known as satellite city.

By late afternoon on April 19, SWBT established a special center for handling service orders for phones needed for the incident area. Every organization within SWBT that played a role in putting in a phone was connected to an open conference line, which was monitored 24-hours a day. When a phone line was requested for a site, each group responded with the required information This system expedited installation and prevented duplicate service orders.

Technical Support Manager Gene Clark and Curtiss Dougherty were in charge of providing telephone service to the incident area. Both men had extensive work in telephone installation and maintenance. Helping them was a core group of employees from the downtown offices and personnel on loan from other city sites and from across the state. The work group numbered 10 to 15 employees each day. The work went on around the clock during the first week of the disaster.

One of the first assignments for Clark and Dougherty was to bring telephone lines to the loading dock on the west end of the Murrah Building, which was being used as a Command Center. To do this, they pulled an inch thick 100-pair conductor cable from the Federal Courthouse on NW 4th Street through the pedestrian tunnel that connected the Courthouse to the Murrah underground parking garage, then through the garage and the Murrah building, and into the loading dock. This line provided connections for about 60 telephones and for computer modems and fax machines needed to support the rescue work. Heavy equipment used in the parking garage damaged the telephone cable that had been placed on the ground. Clark and Dougherty made a number of trips into the garage and eventually attached most of the cable to overhead water pipes to prevent further damage.

SWBT provided over 100 phone lines to the Myriad Convention Center to support the MACC and the USAR teams headquartered there. They provided phone lines for FEMA including a switch for FEMA's communication system. Over 300 lines were installed to support the General Services Administration (GSA) staff at the Medallion Hotel. Temporary lines on 5th Street and on 6th Street served the U.S. Marshal's Command Center, nicknamed "Red October." Temporary lines also served Feed the Children and other agencies and support groups. Four lines were provided to the temporary morgue at the Methodist Church.

The installation work group estimates that roughly 1,500 phones were installed within the perimeter in the first week of the incident. Most of the lines were temporary lines of ordinary telephone wire run from the nearest pole or underground connection., These lines were subject to frequent mishaps. They were pulled down by tall trucks, run over and pulled apart, or damaged by foot and vehicle traffic. The installers were constantly at work repairing, expanding, and improving the system.

Clark and Dougherty used a golf cart to make rounds from One Bell Central through the incident area and to the Myriad Convention Center and back. They stopped along the way to make repairs and take service orders. They carried cellular phone batteries and distributed them upon request. They delivered medical supplies and equipment, loaned out tools, and sometimes brought rescue workers back to One Bell Central.

Several attempts were made to reestablish part of the phone system in the Murrah Building in order to provide better communications for the rescue workers. However, the equipment was so damaged that it was not possible to restore any service.

Southwestern Bell enclosed the covered parking area at One Bell Central with plastic sheeting in order to provide a protected area for rescue workers. Tables were set up in that area and phones provided. Southwestern Bell's personnel began to return to their downtown offices in the week after April 19. Much of their time was given to day-to-day support of the rescue workers who used One Bell Central as a dormitory.

After the first week, requests for new phone services were reduced but the work of caring for all the temporary services was ongoing. Toward May 5, as agencies and some of the media left or reduced their personnel at the site, SWBT began to remove some of the temporary lines.

CELLULAR COMMUNICATIONS

Cellular communications rely on cell sites and mobile switching centers. A large coverage area is divided into smaller areas called cells. In each cell there is a site where radio transmitters communicate with the cellular phones in that cell. Calls travel from the phone to the transmitter in the cell and then to a switching center that sends the phone signal to the number being called. The cellular systems interface with Southwestern Bell's landline phone system.

AT&T Wireless Services and Southwestern Bell Mobile Systems are the two corporate cellular communications providers in Oklahoma City. Both augmented equipment and cellular capacity to handle disaster-related traffic. They provided and distributed hundreds of free phones to all of the government and volunteer agencies involved in the rescue work and provided battery charging and exchange programs. Many persons and agencies in the rescue effort commented on the excellent cellular communications support. On many occasions the two cellular companies worked to support each other in order to provide the best resources to the rescue effort. Their support for this work was accomplished while maintaining operations for the rest of their service areas.

AT&T WIRELESS SERVICES

AT&T Wireless Services operated under the name "Cellular One" during the Murrah bombing incident.

AT&T Wireless Services Sales offices were located at 5509 N. Penn, about five miles north of downtown. The force of the bomb blast was so great it shook the building. The AT&T Wireless Services staff knew a major disaster had occurred and began to react before they knew exactly what had happened. The reports from downtown were broadcast by area radio and television. A meeting of all personnel was called for 9:30 a.m. but was not held. The FBI's headquarters are located in 50 Penn Place. The entire building was evacuated shortly after 9:30 a.m. because of concern that the FBI could be the target for another bomb. For the rest of the day, staff operated from another facility.

AT&T Wireless Services had been an active participant in the Oklahoma Disaster Preparedness Council and had participated in disaster planning with emergency service agencies and area hospitals. They had previously provided backup cellular systems in all area hospitals, the E-9-1-1 Center, and in the Fire Department's Chiefs vehicles.

Bently Alexander, Director of Engineering and Operations for AT&T Wireless Services, provided leadership on the technical side. He was in Seattle on the morning of the bombing and returned to Oklahoma City that day. He was in communication with his staff using cellular phones.

AT&T Wireless Services first response to the bombing was to ascertain the impact of the explosion on its system and network and to determine whether or not any of its equipment had been damaged. None of its facilities or capacities were damaged. The next response was to anticipate and determine what the resulting system impact would be from the demands of the emergency crews, rescue workers, media, and others who would use the cellular system.

For the first 30 minutes, there was a serious problem with congestion as so many people attempted to use their cellular phones. AT&T Wireless Services Engineering and Operations staff knew they needed to balance the system, provide additional capacity, and provide priority phone service to the incident area. AT&T Wireless Services marketing staff knew they needed to implement their disaster plan that called for the immediate provision of phones to the Fire Department and other departments and agencies working at the Murrah site and to develop a plan to distribute and recharge batteries.

The Engineers began making adjustments to the system so that it would handle traffic more efficiently. Four cell sites served the downtown area. They focuses on balancing the traffic so that if one cell was blocked, traffic would move to move to another cell. They also saw the need for a level of priority service for the phones supporting the rescue effort. This could be accomplished by adding a feature to their switch. This was done, and a certain number of channels in each cell site were set aside and assigned priorities within 60 to 90 minutes of the bombing.

The use of the restricted, priority channels prevented congestion and blockage problems on the cellular phones used at the disaster site. For the public at large, non priority phones continued to have some level of blockage until AT&T Wireless Services added additional capacity to the system.

As a part of AT&T Wireless Services preplan with the Oklahoma Disaster Preparedness Council, A Cell on Wheels (COW) was made available for such emergencies. By 9:29 a.m., AT&T Wireless Services engineering personnel had confirmed that the COW was available and ordered it to be mobilized and brought downtown. The best location for the COW would be outside the building at Main and Walker, which housed their downtown switching equipment. Arrangements were made through Lucien Jones of the City to located the COW on Walker Avenue. This required blocking two lanes of traffic to accommodate the cell and its 80 foot tower. The COW was in place and connected to the existing switch, and working by 6:00 p.m. on April 19. It added 30 voice channels to the cellular phone system.

After watching traffic overnight and the next morning, AT&T Wireless Services determined they needed more capacity for the Murrah site. They identified a site a NW 9th and Robinson and entered into handshake agreement with the owner to use the parking lot for the COW and the roof of the building for a microwave and antenna. The began to mobilize the second site at about 11:00 a.m. in the morning and had it operational by 6:00 p.m. The COW was connected to the AT&T Wireless Services switching equipment through a microwave radio. This COW added 38 channels.

The two cells on wheels had added the needed capacity to the system. However, the channels that connected AT&T Wireless Services to landline phones were experiencing blockage. AT&T Wireless Services requested an expedited order for the installation of 48 trunk lines for interconnection to the phone company. This work was done and relieved most of the cellular congestion.

On Friday, April 21, AT&T Wireless Services received a call from GSA regarding communications for the Presidential entourage and Secret Service for the Memorial Service to be held at the Fairgrounds on April 22. In response, a third cell on wheels was mobilized and placed at the fairgrounds and phones provided as needed. This system was put up on the morning of the Memorial Service and taken down that evening.

Also on Friday, AT&T Wireless Services learned that FEMA was setting up a command center at the Myriad and there was discussion that other agencies might be located there. AT&T Wireless Services decided to use a microcell to provide additional phone service at the Myriad. This is a low-power device that has quite a bit of capacity. It would handle the traffic requirements inside the Myriad without increasing the demands on the rest of the system. The vendor for the microcell was AT&T Network Systems. AT&T took a system that was being manufactured, rushed it through completion and testing, and express shipped it to Oklahoma City. AT&T Wireless Services had electricians in the Myriad to provide the necessary power and install the wiring. The equipment arrived at 11:00 a.m. on April 22 and was installed and operational by 4:00 p.m. that afternoon.

The center that FEMA planned for the Myriad had quickly evolved into the MACC. While landline phones were also installed, cellular phones played a major role in communications between agencies at the Myriad and between the MACC and the agencies at the Murrah site.

By the Monday after the bombing, cellular phone traffic was beginning to level out. The three downtown cell on wheels sites and the additional channels that had been added to the existing cell sites had added almost 300 channels to the AT&T Wireless Services System. The capacity was adequate, but there was concern that upward growth could still tax the system. AT&T Wireless Services worked with AT&T to expedite a previous order for planned switch growth, which had been scheduled for July. AT&T expedited the order; however, the new switch was not installed until after the close of the rescue and recovery work at the Murrah Building. The new switch gave AT&T Wireless Services the excess capacity to support disaster deployment in the future, if needed.

While the Engineering and Operations staff worked to provide system capacity, AT&T Wireless Services Marketing staff implemented a plan to provide cellular phones to all agencies at the disaster site. They realized they needed more phones and worked with inventory and district management staff to bring in phones from regional offices and from across the nation.

Key Marketing staff members coordinated immediate distribution of phones, starting with the Fire Department, the FBI, and the ATF. Major Accounts Manager Richard Marley worked with Fire Marshal Gary Curtis to locate a site near the Murrah Building where phones could be distributed and batteries charged. The site at One Bell Central was selected. By 1:00 p.m. on April 19, AT&T Wireless Services began to move a phone supply to that location. Because of security at the disaster site, they ad to park about four blocks away and carry in 30 or 40 phones at a time.

Nine employees staffed the AT&T Wireless Services phone distribution center. Runners were used to take phones and batteries to Command Posts, logistics sites, and non-profit agencies in the disaster area. Phones were also provided to the staff of the Family Assistance program at the First Christian Church. As the rescue and recovery work progressed, AT&T Wireless Services runners familiarized themselves with the various support locations and made rounds with fresh batteries.

They carried messages and helped in every way they could. They were constantly deploying phones and keeping them charged. The distribution center staff and the runners were AT&T Wireless Services workers who volunteered their time to work at the disaster site in addition to regular job duties.

There was a melting pot of various cellular phones allocated to the disaster. Workers kept phone lines open. This depleted the batteries in about an hour. There wasn't the time or the electricity to recharge batteries at the site. AT&T Wireless Services quickly recognized the need to supply one type of phone, especially to those working within the perimeter. They requested all suppliers to provide the Motorola Ultra Classic II with rapid charger. Standardizing the phones as much as possible helped the cellular companies in the job of supporting communications within the perimeter.

The rescue and recovery workers in the Murrah Building were using cellular phones and required a constant supply of new batteries. Someone became concerned that the batteries from the recover site might be a biohazard. The Chief Medical Examiner determined that the batteries were not a biohazard but could be one of the carriers of the flu, stomach viruses and other infections, that were coming into Oklahoma City with all the rescue workers. It was decided to wipe all phones and batteries with disinfectant to minimize the possible spread of infection. AT&T Wireless Services workers did this and wore rubber gloves to protect their hands from the constant use of the disinfectant.

After the opening of the MACC, AT&T Wireless Services moved its distribution site to the Myriad. They set up a post where they could give phones to people as they entered or left. The Myriad provided more electrical power for the constant recharging of the phone batteries. The cellular phones did not have any toll restrictions. Calls made across the country to locate supplies and equipment could be made directly and at no charge.

AT&T Wireless Services estimates that it provided 1,052 phones to the rescue and recovery work and that its contribution of time and equipment was in excess of $4,000,000. AT&T Wireless Services also provided free phones, pagers, chargers, and voice mail to the Oklahoma City Document Management Team that was formed to write the official report of the disaster.

AT&T Wireless Services also provided a specialized camera that was used in the rescue and recovery work. The camera, developed by Dr. Red Duke of Texas, fits on a helmet and transmits a picture every three seconds using cellular transmission. The camera was mounted on a firefighter's helmet and used to provide site information to the structural engineers.

The firefighter could go into a critical area and transmit pictures to the engineers who would determine whether or not it was safe to continue in the area without further stabilization work. In June 1995, the Oklahoma City Rotary Club purchased this piece of equipment and donated it to the Oklahoma City Fire Department.

SOUTHWESTERN BELL MOBILE SYSTEM

Southwestern Bell Mobile Systems (SWBMS) is headquartered at the Oak Tree Center at 9020 N. May Avenue, which is about 10 miles northwest of downtown. The blast was heard and felt at the building. SWBMS responded on two fronts—providing phones and batteries for the rescue work and maintaining and enhancing the cellular communication system.

At the time of bombing, SWBMS had three cell sites with 37 channels serving downtown Oklahoma City. By 9:15 a.m. headquarters knew the source of the blast and ordered a test of the three cell sites. They were determined to be undamaged and in service.

SWBMS personnel closely monitored the traffic on the cell sites serving the downtown area. Most cells in the area overflowed with traffic running five to six times the normal load. About 10 to 15 percent blockage was experienced with calls not going through because of the overload.

By 10:00 a.m., plans were made to add channel capacity to all three faces of the central downtown cell. However, this work could not be performed due to the cell's proximity to the rescue site. SWBMS elected instead to add channel capacity to the cells on the east and west of downtown. They also installed a directed retry program between the downtown cells which allows traffic to overflow another cell face.

At noon the Dallas SWBMS office called to offer use of their cell on wheels (COW). While it appeared that there was enough capacity without this equipment, SWMBS believed they should take every measure to ensure support for the rescue work. The Oklahoma City office accepted the COW and began to make arrangements to place it near the disaster perimeter.

By about 2:15 p.m. on April 19, SWBMS had completed the work to add channels to the west downtown cell site. They had selected a site at NW 7th and Harvey for placement of the COW and made arrangements for electric service and facilities. Arrangements were also made for a crane to be available to hoist the tower section when the COW was delivered.

Technical personnel worked throughout the afternoon. They added six channels to the east cell site so that a total of 49 channels were available to serve the rescue site. Demand for cellular service increased as the media arrived at the disaster site.

SWBMS continued preparations for the installation of the COW, which arrived at 9:30 p.m. John Coles, SWBMS technician from Dallas, supervised installation of the COW. United Communications erected the 40 foot high antenna and installed the coaxial cable that linked the COW to the main telephone switching office. A number of technical difficulties were overcome, and the cell on wheel was operational at 5:00 a.m. on April 20. With the addition of the COW, SWBMS had 65 channels serving downtown. Traffic on the COW was heavy from the moment it went into service.

Monitoring of the system on Thursday afternoon showed that traffic was approaching the Wednesday peak call level. Over 50 percent of the downtown traffic was being handled by the COW and the west cell site. Most of the feedback from the Command Center regarding service quality was positive.

Traffic levels were still high on Friday April 22. The decision was made to add 10 channels to the COW. The channels were added in the afternoon and other technical work performed to reduce possible interference in the rescue area. The additional cells brought the total cells serving the downtown area to 75. Traffic monitoring indicated adequate capacity with the exception of breaking news events. The feedback on call quality and channel feedback was good. SWBMS's technical staff provided continuing support to the system for the remainder of the rescue and recovery effort.

Within 20 minutes of the bombing, SWBMS staff began delivery of phones to emergency personnel. Staff members Chavez Prince, Casey Richards, and Pam Estep coordinated phone distribution and support. They provided a system for 24-hour manning of a battery charging station and rotation of batteries on and off the chargers. Hourly routes were made to deliver fresh batteries at two dozen sites in and around the perimeter. Company personnel volunteered hundreds of hours of time to staff the distribution center. Approximately 500 phones and 850 batteries were provided to city, state, and federal agencies.

In addition to the phones provided to emergency workers, 10 digital cellular phones were provided to key officials. SWBMS had been testing a digital voice system that changes the voice message into binary digital codes that are transmitted to and decoded by the receiving phone. The binary message is scrambled during transmission, providing call security. The phone can also be used as standard cellular phones. The 10 digital phones were tested Wednesday morning and then delivered to Governor Keating, the Attorney General, EMAS, the Oklahoma County Sheriff, the Oklahoma County Emergency Management Agency, and two State Senators.

LONG DISTANCE COMPANIES
Three long distance carriers – AT&T, SPRINT and MCI – provided support for the rescue and criminal investigation work and to the victims and the community.

AMERICAN TELEPHONE AND TELEGRAPH
AT&T provided free long distance service to the MACC and the Urban Search and Rescue task forces at the Myriad Convention Center. This service was used to make calls to locate supplies and equipment needed in the rescue work and allowed Urban Search and Rescue task force members to maintain communications with their families and others in their home cities.

MCI
MCI stationed a satellite truck on Harvey Avenue, adjacent to the One Bell Central command post areas. MCI ran lines from this truck to each Command Post to provide free long distance service. A line was also provided to One Bell Central for free long distance service for the two Urban Search and Rescue task forces based there. On April 22, MCI used a full page ad in the newspaper to offer free long distance services to all its Oklahoma City customers. City residents who were not customers were offered free calling cards which could be obtained at tow locations in the city.

SPRINT
SPRINT provided the toll-free number for the FBI's suspect information hot line. SPRINT provided calling cards for victim's families through the Feed the Children Ministry and the Family assistance Center. Later SPRINT paid the long distance bills for its customers who had family members wounded or killed in the bombing.

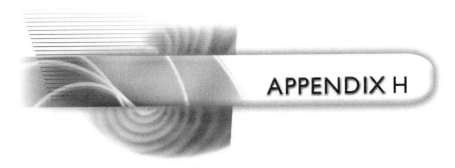

APPENDIX H

The Critical Infrastructure Protection Process Job Aid
Edition 1: May 2002

United States Fire Administration
(Developed by NATEK Incorporated for USFA)

TABLE OF CONTENTS

I. **INTRODUCTION** I
    A. Background 1
    B. Fire and Emergency Medical Community 1
    C. Job Aid Purpose 1

II. **CIP OVERVIEW** 2
    A. Premise 2
    B. Objectives 2
    C. Philosophy 2
    D. Psychology 3
    E. CIP Process Preface 4

III. **CIP PROCESS METHODOLOGY** 5
    A. Identifying Critical Infrastructures 5
    B. Determining the Threat 6
    C. Analyzing the Vulnerabilities 7
    D. Assessing Risk 8
    E. Applying Countermeasures 9

IV. **CIP PROCESS QUESTION NAVIGATOR** 10

V. **INFRASTRUCTURE PROTECTION DECISION MATRIX** 11

VI. **ESTABLISHING A CIP PROGRAM** 12
    A. Justification 12
    B. Program Manager 12
    C. Program Development and Management 12

I. INTRODUCTION

A. BACKGROUND

1. Presidential Decision Directive 63 (PDD 63) was issued May 1998 in response to concerns about potential attacks against critical infrastructures.

2. PDD 63 defined critical infrastructures as the physical and cyber systems so vital to the operations of the United States that their incapacity or destruction would seriously weaken national defense, economic security, or public safety.

3. The directive designated the Federal Emergency Management Agency (FEMA) lead agency for the fire and emergency medical services (EMS) community.

4. FEMA directed the United States Fire Administration (USFA) to increase critical infrastructure protection (CIP) awareness throughout the fire and EMS community.

B. FIRE AND EMERGENCY MEDICAL SERVICES COMMUNITY

1. PDD 63 identified the Emergency Services Sector as one of eight critical infrastructures.

2. The fire and EMS community as well as the law enforcement community comprise the Emergency Services Sector.

3. USFA is the lead critical infrastructure protection (CIP) agency for the fire and EMS community.

C. JOB AID PURPOSE

1. This Job Aid is a guide to assist leaders of the fire and EMS community with the process of critical infrastructure protection.

2. The document intends only to provide a model process or template for the systematic protection of critical infrastructures.

3. It is not a CIP training manual or a complete roadmap of procedures to be strictly followed.

4. The CIP process described in this document can be easily adapted to assist the infrastructure protection objectives of any community, service, agency, or organization.

II. CIP OVERVIEW

A. PREMISE

1. Attacks on the physical and cyber systems of fire and emergency services departments will weaken performance or prevent operations.

2. There are three different types of possible attacks:

   a. **Deliberate** attacks are caused by people (e.g., terrorists, othercriminals, hackers, delinquents, employees, etc.).

   b. **Natural** attacks are caused by nature (e.g., hurricanes, tornadoes, earthquakes, floods, wildfires, etc.).

   c. **Accidental** attacks are caused by HazMat accidents involving nuclear, biological, or chemical substances.

3. These attacks are serious "threats" against critical infrastructures.

B. OBJECTIVES

1. To protect the people, physical entities, and cyber systems that are indispensably necessary for survivability, continuity of operations, and mission success.

2. To deter or mitigate attacks on critical infrastructures by people (e.g., terrorists, hackers, etc.), by nature (e.g., hurricanes, tornadoes, etc.), and by HazMat accidents.

C. PHILOSOPHY

1. Among all the important procedures or things involved in emergency preparedness, CIP is possibly the most essential component.

2. There will probably never be enough resources (i.e., dollars, personnel, time, and materials) to achieve total emergency preparedness.

3. Senior fire and EMS leaders must make tough decisions about what department assets really need protection by the application of scarce resources.

4. There should be no tolerance for waste and misguided spending in the business of emergency preparedness and infrastructure protection.

5. From a municipal perspective, the CIP philosophy is to first protect those infrastructures absolutely required for citizen survivability and continuity of crucial community operations.

6. For the community emergency services, the corresponding CIP philosophy is to first protect those infrastructures absolutely required for the survivability of emergency first responders and the success of their missions.

7. It is impossible to prevent all attacks (e.g., terrorism, natural disasters) against critical infrastructures.

8. CIP can reduce the chances of some future attacks, make it more difficult for the attacks to succeed or degrade infrastructures, and mitigate the outcomes when they do occur.

9. Activities to protect assets essential for the accomplishment of missions affecting life and property are proactive, preemptive, and deterrent in nature, which is exactly what critical infrastructure protection is meant to be.

D. PSYCHOLOGY

1. CIP can be a tool to produce an American "mindset" of protection awareness and confidence in our nation's security and prosperity. Given these new thoughts, it may evoke behaviors that are fully supportive and cooperative with necessary protective measures.

2. CIP may also be a means to change the behavior of terrorists. The proper protection of American critical infrastructures has the potential to develop a new "mindset" among terrorists that their actions will be futile and not yield the results they seek.

3. Community leaders and department chiefs should make occasional public announcements that their critical infrastructures are being protected. This must be done without divulging any details that would be useful to adversaries. Such announcements are not intended to be a ruse or disinformation campaign, but an honest declaration for the "psychological" benefit of both friends and foes.

E. CIP PROCESS PREFACE

1. CIP involves the application of a systematic analytical process fully integrated into all fire and EMS department plans and operations.

2. It is a security related, time efficient, and resource-restrained practice intended to be repeatedly used by department leaders.

3. The CIP process can make a difference only if applied by department leaders, and periodically reapplied when there have been changes in physical entities, cyber systems, or the general environment.

4. It consists of the following five steps:

   a. *Identifying critical infrastructures* essential for the accomplishment of sector missions (e.g., fire suppression, EMS, HazMat, search and rescue, and extrication).

b. **Determining the threat** against those infrastructures.

c. **Analyzing the vulnerabilities** of threatened infrastructures.

d. **Assessing risk** of the degradation or loss of a critical infrastructure.

e. **Applying countermeasures** where risk is unacceptable.

III. CIP PROCESS METHODOLOGY

A. IDENTIFYING CRITICAL INFRASTRUCTURES

1. Identifying critical infrastructures is the first step of the CIP process.

2. The remaining steps of the CIP process cannot be initiated without the accurate identification of a department's critical assets.

3. Critical infrastructures are those physical and cyber assets essential for the accomplishment of missions affecting life and property.

4. They are the people, things, or systems that will seriously degrade or prevent survivability and mission success if not intact and operational.

5. The following are some examples of critical infrastructures:

    a. Firefighters and EMS personnel.

    b. Fire and EMS stations, apparatus, and communications.

    c. Public Safety Answering Points (or 9-1-1 Centers).

    d. Computer-aided dispatch and computer networks.

    e. Pumping stations and water reservoirs for major urban areas.

    f. Major roads and highways serving large population areas.

    g. Bridges and tunnels serving large population areas.

    h. Regional or local medical facilities.

6. Despite many similarities, the differences in physical and cyber systems among individual departments necessitate that senior leaders identify their own critical infrastructures.

7. Remember that protection measures cannot be implemented if what needs protection is unknown!

8. The Fire Department of New York continued to serve the citizens of New York City following the collapse of the World Trade Center towers. However, their ability to do so was tremendously degraded for a period of time given the unprecedented losses of personnel and equipment—the foremost among critical infrastructures.

B.   DETERMINING THE THREAT

1. Determining the threat against identified critical infrastructures is the second step of the CIP process.

2. A threat is the potential for an attack from people, nature, HazMat accident, or a combination of these.

3. The remaining steps of the CIP process depend upon whether or not a department's critical infrastructures are threatened.

4. A determination of credible threat must be made for each critical infrastructure identified in step one.

5. If there is no threat of an attack against one of a department's critical infrastructures, then the CIP process can stop here for that particular asset.

6. If there is only a low threat against one of a department's critical infrastructures (e.g., an earthquake), then leaders can choose to continue the CIP process or stop it here for that particular infrastructure.

7. When there is a credible threat of an attack against a department's critical infrastructures, then it is necessary to determine the following prior to proceeding to the next step of the CIP process:

    a. Exactly which critical infrastructures are threatened?

    b. By whom or what is each of these infrastructures threatened?

8. Two examples of credible threats against critical infrastructures:

    a. "National intelligence assets warn that suspected terrorists may attempt to steal fire trucks or ambulances."

    b. "Police cite increasing incidents of juvenile delinquents breaking into water pumping stations and tampering with equipment."

9. Leaders should concentrate only on those threats that will dangerously degrade or prevent survivability and mission accomplishment.

10. Resources should be applied to protect only those infrastructures for which a credible threat exists!

C.   ANALYZING THE VULNERABILITIES

1. Analyzing the vulnerabilities of credibly threatened infrastructures is the third step of the CIP process.

2. This step requires an examination of the security vulnerabilities (or weaknesses) in each of the threatened infrastructures.

3. A vulnerability is a weakness in a critical infrastructure that renders the infrastructure susceptible to degradation or destruction.

4. There are two types of vulnerabilities to consider in the CIP process:

  a. A weakness in a critical infrastructure that renders the infrastructure susceptible to disruption or loss from a deliberate attack by human adversaries.

  b. A weakness in a critical infrastructure that will further weaken or completely deteriorate as a result of a natural or accidental attack (i.e., natural
disaster or HazMat accident).

5. An efficient vulnerability analysis will examine each credibly threatened infrastructure from the "threat point of view."

6. The analysis will seek to understand the ways by which threats from adversaries, nature, or HazMat accidents might disrupt or destroy the examined infrastructure.

7. If a threatened infrastructure has no vulnerabilities, then the CIP process can
stop here for that particular infrastructure.

8. The CIP process should proceed to the fourth step only for those threatened
infrastructures having vulnerabilities.

9. The following are two examples of vulnerabilities:

  a. Public Safety Answering Points (PSAPs) or 9-1-1

  b. Communication Centers because of their physical locations, power
sources, line routing, Internet-based controls of switching, etc.

  c. Computer Aided Dispatch (CAD) because of its network connections with
Internet connectivity.

10. The protection of threatened and vulnerable infrastructures cannot be accomplished without knowing what or where the vulnerabilities are!

D. ASSESSING RISK

1. Assessing risk of the degradation or loss of a critical infrastructure is the fourth
step of the CIP process.

2. The following priority guidance applies for this assessment:

  a. Threatened and vulnerable infrastructures are a high priority for the application of countermeasures.

  b. Infrastructures that are either threatened or vulnerable, but not both, are a
low priority for protective measures.

3. Focusing on each high priority infrastructure, decision makers must evaluate
the cost of countermeasures in terms of available resources (e.g., personnel,
time, money, materials).

4. The determined costs of protective measures (doing something) for each high
priority infrastructure are now weighed against the impact of the degradation
or loss of that infrastructure (doing nothing).

5. Risk is unacceptable if the impact of the degradation or loss of an infrastructure (doing nothing) is considered catastrophic. The CIP process, therefore, must proceed to the final step for the immediate application of countermeasures.

6. If the impact of the degradation or loss of an infrastructure is not considered great, then decision makers can temporarily decide to accept risk until resources become available.

7. For the infrastructures that are risk adverse and require protection, community leaders should decide the order in which they will receive the allocation of resources and application of countermeasures.

8. For example, research reveals that water pumping stations in rural America are notoriously unprotected. If department leaders follow the CIP process and determine the community pumping station to be a high priority infrastructure, then they should not accept risk and seek local government assistance to apply countermeasures as soon as possible.

9. Failure to assess risk can result in the inefficient application of resources and a subsequent reduction in operational effectiveness!

E. APPLYING COUNTERMEASURES

1. Applying countermeasures where risk is unacceptable is the fifth step of the CIP process.

2. Countermeasures are any protective actions that reduce or prevent the degradation or loss of a critical infrastructure to an identified threat.

3. Countermeasures protect infrastructures and preserve the ability of emergency first responders to efficiently perform their services.

4. They are measures of protection applied to high priority infrastructures that necessitate the allocation of resources.

5. Possible countermeasures differ in terms of feasibility, expense, and effectiveness.

6. Countermeasures can be simple or complex actions limited only by imagination and creativity.

7. In few instances, there may be no effective means to protect a critical infrastructure. Sometimes, prohibitive costs or other factors make the application of countermeasures impossible.

8. Decisions requiring the application of countermeasures will influence personnel, time, and material resources as well as drive the security budget.

9. The following are two examples of countermeasures:

   a. To protect their personnel infrastructure, all FDNY digital radios will be inexpensively reprogrammed so that one channel will override all others and emit a long tone to warn each firefighter to immediately evacuate a building.

      b.   To protect both their personnel and equipment, a growing number of departments are keeping their apparatus bay doors closed at all times.

10.   High priority infrastructures should be considered a loss to plans and operations if not protected by countermeasures!

IV. CIP PROCESS QUESTION NAVIGATOR

DIRECTIONS:

Answer questions for each infrastructure.

- Is the person, thing, or system part of the organization's infrastructure?

- If the answer is **NO**, stop here; but if it is **YES**, then:

- Is this infrastructure essential for survivability and mission success?

- If the answer is **NO**, stop here; but if it is **YES**, then:

- Is there potential for a deliberate, natural, or accidental attack against this critical infrastructure?

- If the answer is **NO**, stop here; but if it is **YES**, then:

- Is the threat of an attack against this critical infrastructure a truly credible one?

- If the answer is **NO**, stop here; but if it is **YES**, then:

- Is there a security vulnerability (or weakness) in the threatened critical infrastructure?

- If the answer is **NO**, stop here; but if it is **YES**, then:

- Does this vulnerability (or weakness) render the critical infrastructure susceptible to disruption or loss?

- If the answer is **NO**, stop here; but if it is **YES**, then:

- Is it acceptable to assume risk and delay the allocation of resources and the application of countermeasures?

- If the answer is **YES**, stop here; but if it is <u>**NO**</u>, then:

- Apply countermeasures to protect this critical infrastructure as soon as available resources permit.

V. INFRASTRUCTURE PROTECTION DECISION MATRIX

DIRECTIONS:

Complete the matrix for each infrastructure.

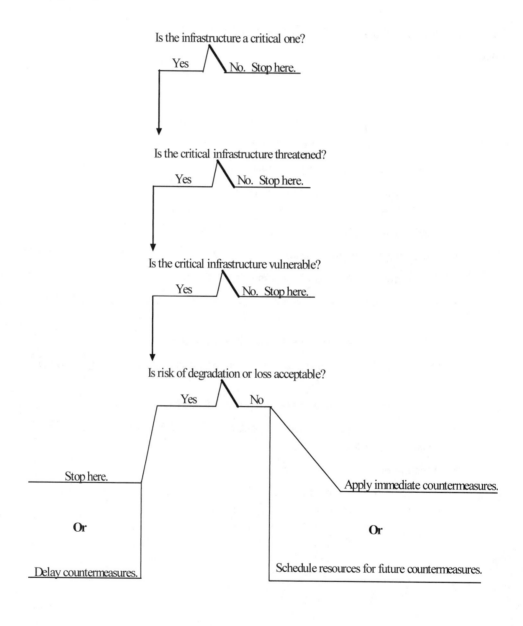

VI. ESTABLISHING A CIP PROGRAM

A. JUSTIFICATION

1. A quality CIP program supports the protection of the people, physical entities, and cyber systems upon which survivability, continuity of operations, and mission accomplishment depend.

2. The terrorist attacks of 11 September 2001 should provide all senior leaders with sufficient justification to immediately implement a critical infrastructure protection (CIP) program within their organizations.

3. If the threat of terrorism itself does not motivate action, then remember that the CIP process also mitigates or eliminates the devastation of critical assets caused by nature and HazMat accidents.

B. PROGRAM MANAGER

1. Critical infrastructure protection is primarily leader business. The department chief, commander, or director appoints a program manager from among the senior leadership of the organization.

2. The program manager administers the CIP program and maintains its value, relevance, and currency.

3. The program manager prepares, obtains approval for, and publishes the program's purpose, strategic goals, and immediate objectives.

4. The program manager proactively initiates actions that protect the organization's critical infrastructures from deliberate, natural, or accidental attacks.

C. PROGRAM DEVELOPMENT AND MANAGEMENT

1. The department chief, commander, or director institutes the organization's CIP program and delegates authority to a manager.

2. The following program development and management steps are recommended:

   a. Select the program manager from among the senior decision-makers of the organization.

   b. Firmly establish the relationship between the organization's mission and the purpose for critical infrastructure protection.

   c. Win support of the department senior and junior leadership, and orient the CIP program to them.

   d. Focus the program on the practice of the CIP process.

   e. After determining which critical infrastructures must receive immediate protection, aggressively seek the resources required to apply countermeasures as soon as possible.

f. Revise and reissue the department security policy to include the CIP Program and the critical infrastructures that demand countermeasures.

g. Brief all department personnel regarding the revised policy and ensure awareness of actions they can take to bolster applied protective measures.

h. Practice operations security (protecting sensitive information) concurrently with CIP.

i. Remain vigilant for threat advisories and new CIP trends, methods, and conditions.

j. Maintain the program by reapplying the CIP process when there have been changes in the physical entities, cyber systems, or the general environment; however, attempt to do so at least semi-annually.

3. The USFA CIPIC will provide assistance (via telephone, electronic mail, or facsimile) to any organization establishing a CIP program. Contact the CIPIC by telephone at 301-447-1325, or by electronic mail at: *usfacipc@fema.gov.* If interested, visit the CIPIC website at: *www.usfa.fema.gov/cipc.*

Critical Infrastructure Protection Information Center
16825 South Seton Avenue
Emmitsburg, MD 21727
(301) 447-1325
www.usfa.fema.gov/cipc
usfacipc@fema.gov

GLOSSARY

Acute exposure. A single encounter with toxic concentrations of a hazardous material or multiple encounters over a short period of time (usually 24 hours).

Administrator password. A secret word or code used to gain access to the highest level of security to administer the computer system.

Air purification devices. Respirators or filtration devices that remove particulate matter, gases, or vapors from the atmosphere. These devices range from full-face piece, dual-cartridge respirators with eye protection to half-mask, mounted-mounted cartridges with no eye protection.

Airways. Any parts of the respiratory tract through which air passes during breathing.

Alkali. A basic substance (pH greater than 7) that has the capacity to neutralize an acid and form a salt.

Alveoli (*singular* alveolus). Microscopic air sacs in which gas exchange between the blood and the lungs occur.

Alveolar ducts. The smallest of the lungs' airways that connect terminal bronchioles and alveolar sacs—sometimes called bronchioles.

Anemia. Any condition in which the number of red blood cells, the amount of hemoglobin, and the volume of packed red blood cells per 100 milliliters of blood are less than normal.

Antidote. An agent that neutralizes a poison or counteracts its effects.

Apnea. Cessation of breathing.

Asphyxia. A condition in which the exchange of oxygen and carbon dioxide in the lungs is absent or impaired.

Bandwidth. The transmission capacity of an electronic line such as a communications network, computer bus, or computer channel.

Biological agent. Living organisms, such as bacteria or viruses, that can cause disease—formally known as pathogens.

Biometrics. Statistical study of biological phenomena.

Caustic. Substance that strongly irritates, burns, corrodes, or destroys living tissue.

Chemical formula. The collection of atomic symbols and numbers that indicates the chemical composition of a pure substance.

Chemical-protective clothing. Clothing specifically designed to protect the skin and eyes from direct chemical contact. Types of chemical-protective apparel include non-encapsulating and encapsulating (referred to as liquid-splash protective clothing and vapor-protective clothing, respectively).

Chronic effect. A pathologic process caused by repeated exposures over a long period of time.

Chronic exposure. Repeated encounters with a hazardous substance over a long period of time.

CCTV. Closed-Circuit Television Systems.

Combustible liquid. Any liquid that has a flash point at or above 100 °F (37.7 °C) and below 200 °F (93.3 °C).

Communications center. A facility either wholly or partially dedicated to being able to receive emergency and nonemergency reports from citizens.

Compressed gas. A gas whose volume has been reduced by pressure.

CAD. Computer-Aided Dispatch assists the telecommunicator in assessing dispatch information and recommends response.

Control zones. Areas at a hazardous materials incident whose boundaries are based on safety and the degree of hazard—generally includes the hot zone, decontamination zone, and support zone.

Corrosive. Ability to destroy the texture or substance of a tissue.

Critical care area. The area in a hospital designated for the treatment of severely ill patients.

CISD. Critical Incident Stress Debriefing.

CISD team. Critical Incident Stress Debriefing team.

Cyanosis. Bluish discoloration of the skin and mucous membranes due to deficient oxygenation of the blood—usually evident when reduced hemoglobin (i.e., hemoglobin unable to carry oxygen) exceeds 5 percent.

Database. A collection of data structures organized in a disciplined fashion in order to obtain information as quick as possible. Databases are made up of two elements—a record and a field.

Decontamination. The process of removing hazardous materials from exposed persons and equipment at a hazardous materials incident.

Decontamination zone. The area surrounding a chemical hazard incident (between the hot zone and the support zone) in which contaminants are removed from exposed victims.

Degradation. The process of decomposition. When applied to protective clothing, a molecular breakdown of material because of chemical contact; degradation is evidenced by visible signs such as charring, shrinking, or dissolving. Testing clothing material for weight changes, thickness changes, and loss of tensile strength will also reveal degradation.

Denial of service. The removal of a LAN service from a user, usually for malicious purposes.

DOH. Department of Health.

Dial-up access. Connection of a device to a network via a modem and a public telephone network.

Dyspnea. Shortness of breath—difficult or labored breathing.

Edema. Accumulation of fluid in body cells or tissues, usually identified as swelling.

Emergency. A sudden and unexpected event requiring immediate remedial action.

EMI. Emergency Management Institute.

Environmental hazard. A condition capable of posing an unreasonable risk to air, water, or soil quality, or plant or animal life.

Epidermis. The outermost layer of the skin.

Explosives. Compounds that are unstable and break down with the sudden release of large amounts of energy.

Explosivity. The characteristic of undergoing very rapid decomposition (or combustion) to release large amounts of energy.

FCC (Federal Communications Commission). Federal organization set up by the Communications Act of 1934 and the revised Telecommunications Act of 1996 that is responsible for nonfederal radio frequency users.

FEMA. Federal Emergency Management Agency.

Firewall. A barrier set up to contain designated LAN traffic within a specified area.

Flame-resistant. Slow or unable to burn.

Flammable. The ability of a substance to ignite and burn.

Flammable (explosive) range. The range of gas or vapor concentration (percentage by volume in air) that will burn or explode if an ignition source is present. Limiting concentrations are commonly called the "lower explosive limit" and "upper explosive limit." Below the lower explosive limit, the mixture is too lean to burn; above the upper explosive limit, the mixture is too rich to burn.

Flash point. The minimum temperature at which a liquid produces enough vapor to ignite.

Flashback. The movement of a flame to a fuel source—typically occurs via the vapor of a highly volatile liquid or by a flammable gas escaping from a cylinder.

Gas. A physical state of matter that has low density and viscosity, can expand and contract greatly in response to changes in temperature and pressure, readily and uniformly distributes itself throughout any container.

Hazard. A circumstance or condition that can cause harm.

Hazardous materials. Substances that, if not properly controlled, pose a risk to people, property, or the environment.

Hazardous materials incident. The uncontrolled release or potential release of a hazardous material from its container into the environment.

HVAC system. Heating, Ventilating, and Air Conditioning system—high voltage.

Hot standby. Backup equipment kept on and running in case some equipment fails.

Hot zone. The area immediately surrounding a chemical hazard incident—such as a spill—in which contamination or other danger exists.

Immediately Dangerous to Life and Health (IDLH). That atmospheric concentration of a chemical that poses an immediate danger to the life or health of a person who is exposed but from which that person could escape without any escape-impairing symptoms or irreversible health effects. A companion measurement to the Permissible Exposure Limit (PEL), IDLH concentrations represent levels at which respiratory protection is required. IDLH is expressed in parts per million (ppm) or mg/m^3.

ICS (Internet Connection Sharing). A way to connect multiple computers in a LAN to the internet through one connection and one IP address.

Incident Commander. The person responsible for establishing and managing the overall operational plan. The Incident Commander (IC) is responsible for developing an effective organizational structure, allocating resources, making appropriate assignments, managing information, and continually attempting to mitigate the incident.

Infrastructure. The base facilities, services, and installations needed for the functioning of a community or society, such as transportation and communications systems, water and power lines, and public institutions including schools, post offices, and prisons.

Interoperability. The ability of software and hardware on multiple machines from multiple vendors to communicate.

JIC. Joint Information Center.

JOC. Joint Operations Center.

Local access. Accessing the computer system without passing through any other networks.

LAN (Local Area Network). A short distance data communications network (typically within a building or campus) used to link together computers and peripheral devices (such as printers) under some form of standard control.

Main server. A computer system in a network that is shared by multiple users.

MMRS. Metropolitan Medical Response.

Mist. Liquid droplets dispersed in air.

NMRT. National Medical Response Team.

NTIA (National Telecommunications and Information Administration). Agency of the United States Department of Commerce responsible for development of communication (primarily telephone) standards.

Network. Connection for all types of computers and computer-related things (ie. terminals, printers, modems, etc.)

Network bandwidth. The amount of data that can be sent through a network connection measured in bits per second.

PSWAC. Public Safety Wireless Advisory Committee.

Physical state. The solid, liquid, or gas of a chemical under specific conditions of temperature and pressure.

PDD 63. Presidential Decision Directive 63 provides the basis for federal action and serves as an appeal to the private sector to be its partner in protecting United States infrastructures.

PSAP. Primary Safety Answering Point.

Prophylaxis. Prevention from or a protective treatment for a disease.

PIO. Public Information Officer.

Reactivity. The ability of a substance to chemically interact with other substances.

Redundant system. One or more "backup" systems available in case of failure of the main system.

Rescuer protection equipment. Gear necessary to prevent injury to workers responding to chemical incidents.

Response organization. An organization prepared to provide assistance in an emergency (e.g., fire department).

Response personnel. Staff attached to a response organization (e.g., hazmat team).

Routes of exposure. The manner in which a chemical contaminant enters the body (e.g., inhalation, ingestion).

Secondary contamination. Transfer of a harmful substance from one body (primary body) to another (secondary body), thus potentially permitting adverse effects to the secondary body.

Self-Contained Breathing Apparatus (SCBA). Protective equipment consisting of an enclosed facepiece and an independent, individual supply (tank) of air—used for breathing in atmospheres containing toxic substances or underwater.

Smart gateways. A computer that routes traffic from a workstation to an outside network that is serving web pages. The gateway will be programmed with information allowing trusted traffic to pass through and stopping untrusted traffic. It is up to the user to determine what constitutes trusted and untrusted traffic.

Soluble. Capable of being dissolved.

Solution. A homogeneous mixture of two or more substances, usually liquid.

Solvent. A substance that dissolves another substance.

SOP. Standard Operating Procedures.

Support zone. That area beyond the decontamination zone that surrounds a chemical hazard incident in which medical care can be freely administered to stabilize a victim.

Talk groups. A unique number representing a group of radio users in a radio system.

Toxic. Having the ability to harm the body, especially by chemical means.

Toxic potential. The inherent ability of a substance to cause harm.

USAMRI. United States Army Medical Research Institute.

USFA. United States Fire Administration.

Upgrade. A newer version of a software or hardware product designed to replace an older version of the same product.

Vapor. The gaseous form of a substance that is normally a solid or liquid at room temperature and pressure.

Vapor density. The weight of a given volume of vapor or gas compared to the weight of an equal volume of dry air, both measured at the same temperature and pressure.

Vapor pressure. A measure of the tendency of a liquid to become a gas at a given temperature.

VHF (Very High Frequency). The portion of the electromagnetic spectrum with frequencies between 30 and 300 MHZ.

VPN (Virtual Private Network). Carrier-provided service in which the public switched network provides capabilities similar to those of private lines.

Virus. A software program capable of duplicating itself and usually capable of wreaking great harm on a computer system.

Virus protection software. A program that searches a hard drive for viruses and removes any that is found.

Water-reactive material. A substance that readily reacts with water or decomposes in the presence of water, typically with substantial energy release.

INDEX

A

AAS-PT (Associate in Applied Sciences-Public Safety Telecommunicator) 73
abortion 9
Abu Nidal Organization 10
access to buildings 90
accidents 20
Action Direct 8
ADL (Anti-Defamation League) 210
administrator access 117
aerial measuring system 62
aerosolization 192–193
Afghanis (mercenaries) 8–9
Afghanistan 8–9
Africa 8
AIDS 17
AIDS virus 176
air locks 89
airlines 2, 25
airports 84
alarm systems 88
alerting and warning systems 36
Alfred P. Murrah Federal Building bombing 2
   communications recommendations 231–232
   communications units (911) 229–231
   community actions 229–251
   domestic terrorism in 9
   and facility security issues 85
   Oklahoma City communications 233–251
      cellular communications 245–251
      city radio channels 235
      corporate communication systems 242–245

      Fire Communication Maintenance Center 234–235
      Fire Dispatch Center 235–237
      government communication systems 240–242
      long distance carriers 251
      mobile data terminals 240
      Police Communications Center 237–240
      two-way radio system 233
American Red Cross 28, 53, 211
aminoglycosides 204
anarchists 3
ANG (Army National Guard) 50
animal rights 9
annual development increment 30
annual work increment 31
antennas 89
anthrax 178
   decontamination and isolation 200
   description of 199
   diagnosis of 199
   handling 191–193
   Internet resources 213
   in letters or packages 191–193
   outbreak control 200
   prophylaxis 200
   signs and symptoms 199
   treatment of 199–200
   See also biological agents
antibiotics 187
Anti-Defamation League (ADL) 210
Arab governments 9
Army National Guard (ANG) 50
Aryan Nation 177

Associate in Applied Sciences-Public Safety Telecommunicator (AAS-PST) 73
Association of Public Safety Communications Officials (APCO) 12, 207
AT&T Wireless Services 245–249
audio equipment 145
audiocassette tapes 145
Aum Shinrikyo 8
avalanches 19, 21
Azerbaijan 7

B

Bacillus anthracis 199
backdraft 18
backup communications centers 124
backup radio system 171
backups of data 124
Bagwhan, Rajineesh 181
ballistic glass 88
Bangladesh 7
bank robberies 176
barriers 85
batteries 145
Bin Laden, Osama 174
biological agents 178
    anthrax 199–200
    brucellosis 200–202
    contamination by aerolization 192–193
    exposures to 187–188
    first-responder actions to 190
    indicators of 181, 195–196
    Internet resources 213–214
    notification essentials 196
    plague 202–203
    proliferation of 21
    smallpox 204–205
    sources of 175
    suspected release of 196
    tularemia 203–204
    See also NBC (nuclear, biological, or chemical) agents/events
biological clouds 89
biological terrorism 21

counter-terrorism training 76–77
decontamination in 183–184
first responder concerns 181–183
indications of 181
initial actions by dispatch personnel to 183
initial actions by first responders to 183
patient management in 184–188
Web sites 211–212
    See also terrorism
biometrics 90–92
BL-3 containment conditions 204
blast films 85–88
B-NICE (biological, nuclear, incendiary, chemical, and explosive) attacks 73
bomb incident plans 93–94, 95
    command center 99–100
    evacuation plans in 96
    responding to bomb threats 95
    search teams 96–99
    security agaisnt bomb incidents 95
    steps in 99
bomb threats 95
bombs 190
boot disks 146
Bosnia 9
botulinum antitoxin 187
Branch Davidian 8, 177
Brian (typhoon) 19
bridges (computers) 122
Brigate Rosse 8
Brucella 200
brucellosis 200–202
buboes 202
bubonic plague 202
bug bombs 159
building codes 84
buildings, access to 90
Burma 7
Burundi 8

C

cables and wires 117
CAD (computer-aided dispatch) system 229

call handling 25
call-takers 178
camcorder policy process 22
cameras 85, 164
Canadian Security Intelligence Service 209
capability assessment 29, 34–37
capability maintenance 29
capability shortfall 30
car bombs 85
caregivers 137–139
Carlos the Jackal 10
casualties 195
cellular communications 245–251
Cellular One 245
cellular phones 146
Cellules Communistes Combattants 8
CEM (Comprehensive Emergency Management) 24, 25–27
Center for Democracy and Technology (CDT) 209
Center for Nonproliferation Studies (CNS) 216
Center for Research on the Epidemiology of Disaster (CRED) 215
center line staff 173
Centers for Disease Control and Prevention (CDC) 207
    role in nuclear events 50
    Web site 211
Central Intelligence Agency (CIA)
    indirect support for Mojahedin groups in 1980s 9
    Web site 208
Chechnya 7
chemical incidents 189–190
chemical warfare (CW) agents 21, 178
    indicators of 195
    notification essentials 196
    sources of 175
    suspected release of 196
    See also NBC (nuclear, biological, or chemical) agents/events
Chemical/Biological Incident Response Force (CBIRF) 49
Chemical/Biological Rapid Response Team 58

Chicago tunnel flood 19
China 9
chloramphenicol 202
chordopoxvirus 204
Christian extremist groups 9
ciprofloxacin 199, 200
CISD (Critical Incident Stress Debriefing) 131–135
civil defense. See emergency management
civil disorders 21
closed circuit television systems 90
clouds, as indicator of chemical warfare agent usage 195
CNN syndrome 22
Cold War, end of 7, 21
Colombia 8
command centers, in bomb security plans 99–100
Command, Control, and Communications (C3) 111
communications center director 117, 161
communications centers 24–25
    bomb incident plans 93–94
        command center 99–100
        evacuation plans 96
        responding to bomb threats 95
        search teams 96–99
        security against bomb incidents 95
        steps in 99
    communications tower security 100–101
    computer security 115–125
    contingency policies 169
    dealing with news media 169
    emergency planning and drills in 100
    evacuation procedure 221–228
    facility security 83–84
        access to buildings 90
        access to communications rooms 92–93
        biometrics in 90–92
        building perimeter 85–89
        evaluating 162–163
        outside facilities 85
        planning 164–166

communications centers, facility security (*cont.*)
   signature metrics in 92
   surveying 84–87
   training 167–168
   utillities and ventilation systems
      89–90
   voice analysis in 92
handling strangers in 163–164
new building construction in 100
personnel management
   caregivers 137–139
   committee setup 131
   Critical Incident Stress Debriefing
      (CISD) 131–135
   during terrorist events 127–131
   recovery plans 136–137
Public Information Officers (PIOs)
   142–143
risks for terrorist attacks 84
security policies 162
and terrorism preparedness plans 57
terrorism training programs 67–69
   availability of 69
   colleges and universities 73
   government-sponsored 74–78
   Helpline 78–80
   managing 70–71
   National Fire Academy courses
      72–73
   post-training 80–81
   self-study courses 71–72
   videotape course 77–78
as first responders in terrorism attacks
   71
communications operators. See telecommunicators
Comprehensive Emergency Management
   (CEM) 24, 25–27
computer center managers 118
computer hardware technicians 118
computer systems 115–116
   access to 117–119
   authorized personnel 117–119
   in backup communications centers 124
   cables and wires 117

data backup 124
dial-up access 119–120
direct connection to Internet 120–121
firewalls 121
and increase in call volume 125
indirect connection to Internet 122
in Joint Information Centers 146
local access to 119
manual operation of 124–125
physical security 116–117
redundant data 124
redundant systems 123–124
security risks 122–123
virtual private networks (VPNs) 121
viruses 125
computer-aided dispatch (CAD) system 229
concrete barriers 85
Constitution 104
consular information sheets 208
consumer terrorism 9
contamination 184, 192–193
contigency policies 169
convenience stores 85
corporate communication system 242–245
Corps of Engineers 53
Corsican separatists 7
counterterrorism 207, 210–211
counter-terrorism training courses 76–77
counties 104
courthouses 84
cowpox 205
crisis and consequence management 44–45
crisis situations 105
Critical Incident Stress Debriefing (CISD)
   131–135
critical infrastructure protection 212–213
   decision matrix 262
   establishing program for 263–264
   Job Aid 253–264
   methodology 257–261
   overview 255–257
Cuba 177
cults 178
cyanide 178
cyanosis 199

D

dam failures 19
data backup 124
database program 146
databases, monitoring of 116
dead animals 195
decontamination 186–187
    of anthrax 200
    in biological terrorism events 183–184
    of plague 203
    of smallpox 205
    of tularemia 204
Defense Civil Preparedness Agency 23
Defense Threat Reduction Agency
    (DTRA) 215
Department of Agriculture 53
Department of Defense (DOD)
    Chemical/Biological Rapid Response
        Team 58
    Joint Task Force for Civil Support 57
    role in Federal Response Plan 49
    role in NBC incidents 57
    Web site 212
Department of Energy (DOE) 50, 54, 62
Department of Health and Human Services
    53, 60–61
Department of Justice (DOJ) 78, 208
Department of Transportation (DOT) 53, 63
Department of Veteran Affairs 63
dial-up access 119–120
diaphoresis 199
Disaster Center 215
disaster management. See emergency man-
    agement
Disaster Medical Assistance Teams
    (DMATs) 49
Disaster Mortuary Operational Response
    Teams 61
discharge management 186
dispatch personnel 183
ditches 85
DMATs (Disaster Medical Assistance
    Teams 49

domestic terrorists 3
doomsday cults 178
doxycycline 200, 201, 202
droughts 19, 20
dyspnea 199

E

Earth Liberation Front (ELF) 177
earthquakes 19–21
East-West politics 7
economic terrorism 2
edema 199
Edgewood Research, Development, and
    Engineering Center (ERDEC) 49
Electronic Frontier Foundation 210
Electronic Privacy Information Center 209
e-mail 125
Emergency Broadcast System (EBS) 36, 144
emergency communications 35
emergency generators 89
emergency management 16
    changing context of 21–22
    and communications centers 24–25
    comprehensive 25–27
    effects of news media on 22
    emergency managers 22–23
    exercising 40–41
    factors in 17–19
    fire services in 16, 16–19
    and potential hazards 20–21
emergency management agencies 15
Emergency Management Institute 142
emergency management organizations 35
emergency management systems
    benefits to communities 17
    developing 16
    integrated. See Integrated Manage-
        ment Systems (IEMS)
    plan maintenance program 40–41
emergency managers 22–23
Emergency Medical System (EMS) 45
Emergency Operations Center (EOC)
    46–47, 231

Emergency Operations Plan (EOP)
    and capability assessment 35
    components of 51–52
    developing 29
    exercises 52
    exercising 40–41
    organizing 38
    setting goals in 37
    Web sites 216–217
    writing 51–52
emergency power 36
emergency public information 36
emergency reporting 37
Emergency Response and Research Institute (ERRI) 210
Emergency Response Team (ERT) 65, 230
Emergency Response to Terrorism: Job Aid (ERT:JA) 71–72
Emergency Response to Terrorism: Self Study (ERT:SS) (Q534) (course) 71
Emergency Support Functions (ESFs) 46–47, 52–54
emergency support services 37
enemy, as a factor in terrorism 5
Energetic Materials Research and Testing Center 211
energy 54
ENRON, collapse of 2
environmental groups 9
Environmental Protection Agency (EPA) 50, 64
ERDEC (Edgewood Research, Development, and Engineering Center) 49
ETA (Euzkadi Ta Askatasuna) 7
ethnic groups 8
Europe, terrorist attacks against US tourists in 2
Euro-terrorists 8
Euzkadi Ta Askatasuna (ETA) 7
evacuation 37
    and bomb threats 96
    pre-evacuation issues 221
    standard operating procedure 219–228
    types of 220–221
exposures to biological agents 187–188
extremist viewpoints 4

F

facial recognition 92
facility security 83–84
    access to buildings 90
    access to communications rooms 92–93
    biometrics in 90–92
    bomb incident plans 93–94
        command center 99–100
        evacuation plans 96
        responding to bomb threats 95
        search teams 96–99
        security against bomb incidents 95
        steps in 99
    breaches in 171
    building perimeter 85–89
    communications tower security 100–101
    emergency planning and drills in 100
    evaluating 162–163
    and new building construction 100
    outside facilities 85
    planning 164–166
    signature metrics in 92
    surveying 84–87
    training 167–168
    utilities and ventilation systems 89–90
    voice analysis in 92
fax machines 145–146
fear as agent of change 2
Federal Aviation Administration (FAA) 209
Federal Bureau of Investigation (FBI) 44
    authority over NBC (nuclear, biological, or chemical) events 47–48
    Web site 208
Federal Communications Commission (FCC) 107
Federal Coordinating Officer (FCO) 151
Federal Emergency Management Agency (FEMA) 12
    crisis and consequence management 44
    Emergency Response Team (ERT) 65
    headquarters-level response structure 150–151

initial and continuing actions 153
participation in Joint Information Centers (JICs) 147–150
regional-level response structure 151
responsibilities in NBC events 50
Federal Radiological Monitoring and Assessment Center 62
Federal Response Plan (FRP) 48–49
emergency support functions (ESFs) in 52–54
Web site 211
filtration systems 89
fingerprint recognition 91
Fire Communications Center (Oklahoma City) 234–235
Fire Dispatch Center (Oklahoma City) 235–237
fire fighting 53
defensive mode in 18
fatalities in 17
hazards in 18
offensive modes in 18
fire incidents, costs of 17
fire services 16–19, 23
Firefighting Resources of Southern California Organized for Potential Emergencies (FIRESCOPE) 39–40
Fireground Command System 39
firewalls (computers) 121
first responders 183
golden rule for 189
in terrorist incidents 189
flashover 18
floods 19, 20
fog 195
foggers 159–160
food 54
Food and Nutrition Service 54
formatted disks 146
Francisella tularensis 203
Front de la Liberation Nationale de la Corse (FLNC) 7
fuel supplies 54
fumes 89

G

garbage containers 85
gateways 122
General Services Administration (GSA) 53
generators 89
gentamicin 201, 204
geological events 19
Georgia 7
glass film 88
government communication systems 240–242
Government Printing Office (GPO) 72
guerrilla warfare 7

H

hackers 120–121
Hamas 7
handprint recognition 91
hardware technicians 118
Harrods bombing 9
harware vendors 118–119
hate groups 177
Hawaii 19
hazard analysis 29, 32–33
key terms in 33–34
planning process in 33
hazardous materials
in Federal Response Plan 53
incidents 20–21
health surveillance 48
heat detectors 85
Hellenic Resources Institute 207
Helpline 78–80
hepatitis 17
highjacking of airlines 25
high-risk terrorist groups 10
highway terrorism 2
holy war 9
homeland security 104
hotels 131
Hurricane Andrew 19
Hurricane Iniki 19

hurricanes 19–21
HVAC systems 89–90
hydraulic barricades 85

I

identification cards 91
immediate evacuation 221
Immigration and Naturalization Service 209
imminent evacuation 221
immune globulin 187
immunization 187
Incident Action Plan (IAP) 45
Incident Command Post (ICP) 50
Incident Command System (ICS) 28
    functions 45
    history of 38–40
    integration with federal government 47–50
    interface with emergency operation centers 43–65
Incident Commander (IC) 18
Incident Management Systems (IMS) 45
incomplete evacuation 221
India 9
India, 7
infrared camera scanning 91
in-hospital care 48
insect foggers 159–160
insurgencies 7
Integrated Emergency Management System (IEMS) 27
    capability assessment 34–37
    concept 28
    hazard analysis 32–33
    overview 27
    process 28–31
International Association of Counterterrorism and Security Professionals (IACSP) 209
International Association of Emergency Managers (IAEM) 216
International Critical Incident Stress Foundation (ICISF) 216

international terrorism 3, 7, 177
Internet 122
    direct connection to 120–121
    and hate crimes 177
    indirect connection to 122
interoperability 107
    channels 113
    planning 108–110
    and public safety services 107–108
    technology 111–113
intolerance, as a factor in terrorism 4
iodine 200
Iran 9, 177
Iraq 177
iris scanning 91
Islamic groups 8–9

J

jails 84
Jammu and Kashmir Liberation Front (JKLF) 9
Japanese Red Army 10
jihad 9
Job Aid 71–72, 253–264
Joint Information Centers (JICs) 141–145
    continuing actions 153–160
    equipment 145–147
    headquarters-level response structure 150–151
    regional-level response structure 151
    response actions 153
    role of federal government in 147–150

K

Kaczynski, Theodore (Unabomber) 10
Kashmir 9
key cards 116
Kim-spy (Web site) 210
knee walls 85
Kosovo 8
Ku Klux Klan 9

L

landslides 19, 21
Lebanon 9
left-wing groups 3, 8
letters, suspicious 191–193
Libya 177
lighting 85
line supervisors 171–172
local access, to computer systems 119
local building codes 84
local governments 31
local terrorists 177
locks 171
lone individuals as terrorists 177
long distance carriers 251
Los Angeles riots 19
low-risk groups 11

M

Macedonia 8
man-made threats 19
Marshals Service 208
Marx, Karl 3
Marxism 3, 8
mass care 53
MCI 251
McVeigh, Timothy 2
media and emergency management 22
medium risk terrorist groups 10–11
meningitis 202
mental health 49
mercenaries 9, 10
messianic terrorism 8
meteorological events 19
Metropolitan Medical Response System (MMRS) 56
Metropolitan Medical Strike Team (MMRS) 47
Microsoft Word (software) 146
Middle East 9
militia groups 9
minimal-risk terrorist groups 11

mitigation 26, 30
mobile data terminals 240
Mobile Emergency Response System (MERS) 232
modems 119–120, 147
Mojahedin groups 8
monkeypox 205
movement detectors 85
multi-Year development plans 30
Murrah Federal Building bombing. See Alfred P. Murrah Federal Building bombing
Muslim Kashmiris 8
Myanmar 7

N

narcissistic terrorists 10
National Communications System 53
National Disaster Medical System (NDMS) 49
National Emergency Number Association (NENA) 12, 207
National Fire Academy (NFA) 40, 72
National Fire Protection Association (NFPA) 39
National Governors Association 23
National Institute of Justice (NIJ) 211
National Pharmaceutical Stockpile 61
National Security Agency (NSA) 209
National Telecommunications and Information Administration (NTIA) 107
nationalist groups 7–8
natural disasters 19
natural threats 19
NBC (nuclear, biological, or chemical) agents 21
    common agents 178
    sources of 175
NBC (nuclear, biological, or chemical) events 43–65
    areas of competency in handling 74–75
    counter-terrorism training 76–77
    essential notifications in 196
    full authority of FBI over 47–48

NBC (*cont.*)
  incident objectives 196
  specialized agencies in 49
neo-fascists 3
nerve agents 190
new building construction 100
New Mexico Institute of Mining and
  Technology 211
news media and emergency management
  22, 169
NFA (National Fire Academy) 40, 72
NFPA (National Fire Protection
  Association) 39
911 emergency lines 229–231
non-aligned groups 177
North Africa 9
North Korea 177
notifications in NBC (nuclear, biological, or
  chemical) events 196
NTIA (National Telecommunications and
  Information Administration) 107
nuclear attacks 21
nuclear incidents 190
Nuclear Regulatory Commission 65
Nunn-Lugar-Domenici (NLD) Domestic
  Preparedness Program 78–79

O

oceanic events 19
Office for Domestic Preparedness (ODP)
  78–79
Office of Homeland Security 11
Office of Justice Programs (OJP) 78
ofloxacin 201
Oklahoma City bombing. See Alfred P.
  Murrah Federal Building bombing
Oklahoma State Bureau of Investigations
  (OSBI) 239
Omar (typhoon) 19
Orange County Fire Rescue Communica-
  tions Center 219–228
outbreak control
  anthrax 200

brucellosis 201
plague 203
smallpox 205
tularemia 204
See also biological agents

P

packages, suspicious 191–193, 195
padlocks 171
paging systems 169
Pakistani Intelligence 9
Palestinian groups 7
parks department 23
passwords 91
patient management 184
  and decontamination of environment
    186–187
  discharge management 186
  isolation precautions in 185
  patient placement 186
  patient transport 186
  prophylaxis 187
  psot-exposure immunization 187
  and pyschological aspects of WMD
    (weapons of mass destruction) 188
  triage 187–188
patriot groups 177
penicillin 199
pepper spray 178
personal identification numbers (PINs) 91
personal threats 176
personnel management
  caregivers in 137–139
  committee setup 131
  Critical Incident Stress Debriefing
    (CISD) 131–135
  during terrorist events 127–131
  recovery plans 136–137
pharmaceuticals 61
Philippines 9
photoelectric cells 85
PIOs (Public Information Officers)
  142–143

PIRA (Provisional Irish Republican Army) 7, 9
plague 178
    decontamination and isolation 203
    description of 202
    diagnosis of 202
    Internet resources 214
    outbreak control 203
    prophylaxis 202–203
    signs and symptoms 202
    treatment of 202
    See also biological agents
planned evacuation 220
Planned Parenthood 176
planters 85
Police Communications Center (Oklahoma City) 237–240
politicos 8
pollutants 89
portable audiotape recorders 145
Postal Inspection Service 209
post-exposure immunization 187
power failures 20
power systems 54
pre-evacuation issues 221
pre-event recognition of terrorist events 5–6
preparedness 26
Presidential Decision Directive 39 (PDD-39) 44
Presidential Decision Directive 63 (PDD 63) 254
primary safety answering points (PSAPs) 129
property loss from fires 17
prophylaxis 187
    anthrax 200
    brucellosis 201
    plague 202–203
    smallpox 205
    tularemia 204
    See also patient management
Provisional Irish Republican Army (PIRA) 7, 9
PSWAC (Public Safety Wireless Advisory Committee) 107

public health 37, 49
public information 36
Public Information Officers (PIOs) 142–143
public safety 103–107
public safety answering points (PSAPs) 25
public safety services 107–108
public safety support providers 108
Public Safety Wireless Advisory Committee (PSWAC) 107
public services 108
public shelters 36
public works 23, 37, 53

R

Radiation Emergency Assistance Center 62
radio interoperability 107
    channels 113
    planning 108–110
    and public safety 107–108
    technology 111–113
radio system 171
Radiological Advisory Medical Team 59
radiological assistance program teams 61
radiological incidents 21, 190
rail transportation incidents 21
Rapid Response Information System 76
rebellion 7
recovery 26
Red Cross 28, 53, 211
redundant data 124
redundant systems 123–124
religious extremist groups 8–9
religious radicalism 8
repair vendors 162
response activities 26
response phase (terrorist events) 6
restaurants 131
retina scanning 91
revolutions 7
ricin 178
rifampin 201
rights, protection of 104

right-wing groups 3, 9
riots 19
risk analysis 32–33
rodent control 203
Rote Armee Fraktion 8
Rumor Control Center 143–144
rural fires 53
Russian Army 7
Rwanda 8

S

salmonella bacteria 181
Salvation Army 28
San Andreas Fault 19
SARA (Superfund Amendments and Reau-
    thorization Act) 41
sarin gas 178
Saudi Arabia 9
SBCCOM (Soldier and Biological Chemi-
    cal Command) 74
screening 90
scrubbers 89
search teams 96–99
Secret Service 50
security breaches 171
security cameras 85, 164
security policies 162
seismic events 19
self-study courses, on terrorism emergency
    response 71–72
September 11, 2001 attacks 19
    and emergency response systems 69
    and employee stress 128
    as wake-up call in fighting terrorism 179
    economic cost of 19
    and public safety 103
septicemia 199
serology 204
sewer system 163
shelters 36
sheriff 104
signature metrics 92
Sikh Punjabis 8

single-issue groups 3, 9–10
smallpox 178
    decontamination 205
    description of 204–205
    diagnosis of 205
    Internet resources 213
    outbreak control 205
    prophylaxis 205
    signs and symptoms 205
    sources of 176
    treatment of 205
    See also biological agents
smoke 89
social threats 20
sodium hypochlorite 200
software vendors 118–119
Soldier and Biological Chemical Command
    (SBCCOM) 74
solid waste disposal 49
Somalia 9, 205
Southwestern Bell Mobile Systems 249–251
Southwestern Bell Telephone Co. 242–245
Soviet Union, collapse of 21, 175
Special Events Contingency Planning for Pub-
    lic Safety Agencies (course) 71
Special Medical Augmentation Response
    Team 58–59
special-interest terrorists 3
speed bumps 85
SPRINT 251
Sri Lanka 7, 8
state emergency operation plans 216–217
state emergency service agencies 56
state governments 31
state police 104
strangers 163–164
streptomycin 201, 202, 204
stress 128
    and personal needs of communications
        operators 129–131
    reactions to 133–135
    sources of 129
stridor 199
structural fires 20
subconflict organizations 8

subsidences 21
Sudan 9, 177
Superfund Amendments and Reauthorization Act (SARA) 41
surrogate terrorism 10
suspicious letters and packages 195
    handling of 191–193
Syria 177
system administrators 117

T

Tajikistan 9
talk groups 113
targets of terrorism 3, 5
Technical Escort Unit (TEU) 49, 58
technical writing 51
technicians 170–171
telecommunications 53
telecommunications failures 20
telecommunicators
    caregivers 137–139
    committee setup 131
    Critical Incident Stress Debriefing (CISD) of 131–133
    personal needs during terrorist events 127–131
    role in communications recovery plans 136–137
    source of stress 129
    stress reactions 133–135
    terrorism training programs 67–69
        availability of 69
        colleges and universities 73
        government-sponsored 74–78
        Helpline 78–80
        managing 70–71
        National Fire Academy courses 72–73
        post-training 80–81
        self-study courses 71–72
        videotape course 77–78
terrorism 176
    commonly accepted variables of 2
    consumer 9
    countries that sponsor 177
    definition of 1–2
    fear as agent of change in 2
    first responders to 189
    intended audience of 3
    international 7
    Internet resources 207–208
    as man-made threats 20
    messianic 8
    and personnel management 127–131
    and public safety 103–106
    recognition of 5–6
    surrogate 10
    and WMD (weapons of mass destruction) 174
terrorism preparedness plans 43–65
Terrorism Research Institute 209
terrorism training programs 67–69
    availability of 69
    college and universities 73
    government-sponsored 74–78
    Helpline 78–80
    managing 70–71
    National Fire Academy courses 72–73
    post-training 80–81
    self-study courses 71–72
    videotape course 77–78
terrorist events 5–6
    pre-event recognition of 5–6
    response phase 6
terrorists and terrorist groups 3
    categories of risks of 10–11
    desired outcome of 3
    local 177
    major beliefs of 3–5
    preference for bombs and conventional explosives 84
    targets of 3, 5
    types of 3
        doomsday cults 178
        hate groups 177
        international 177
        left-wing 8
        nationalists 7–8

terrorists and terrorist groups, types of (*cont.*)
    patriot groups 177
    religious extremist groups 8–9
    single-issue groups 9–10
    surrogate groups 10
  victims of 3
tetracycline 204
thermal recognition 92
thermonuclear weapons 21
threats 19, 173–176
  as incident to another crimes 176
  low-profile 21
  man-made 19
  natural 19
  personal 176
  ranking of 20
  suspects 177–178
  terrorist acts 176
thunderstorms 19
tick bites 204
timers 85
tornadoes 19, 20
tourism industry, effect of terrorism on 2
training 70
transmission equipment, 89
transportation accidents 20
traps 85
trash bins 85
travel advisories 208
travel industries 2
triage 187–188
tropical storms 19–21
truck bombs 85
trunked radio systems 113
tsunamis 19, 21
tularemia 203–204
tularensis 203
two-way radio systems 111
typhoons 19

U

U.S. Army Chemical and Biological Defense Command 78

U.S. Army Corps of Engineers 53
U.S. Army Medical Research Institute of Infectious Diseases 216
U.S. Army Radiological Advisory Medical Team 59
U.S. Army Soldier and Biological Chemical Command 74
U.S. Army Special Medical Augmentation Response Team 58–59
U.S. Army Technical Escort Unit 58
U.S. Coast Guard 63
U.S. Department of State 208
U.S. Forest Service 53
U.S. Marine Corps Chemical-Biological Incident Response Force 59
U.S. Marshals Service 208
U.S. Office of Homeland Security 11
U.S. Postal Inspection Service 209
U.S. Public Health Service 53
U.S. Public Health Service (USPHS) 46
U.S. Secret Service 50
UHF frequency bands 111
Unabomber 10
unexplained casualties 195
United States Constitution 104
unpasteurized dairy products 200
urban fires 20, 53
Urban Search and Rescue (USAR) 53, 56
user access to computer systems 118
utilities 89–90

V

vaccines 187
vaccinia virus 205
varicella virus 205
variola virus 205
vector control 49
vendor access to computer systems 118–119
ventilation systems 89–90
very-high risk terrorist groups 10
VHF frequency bands 111
victims of terrorism 3
video equipment 145

video monitors 116
violence, threat of 2
virtual private networks (VPNs) 121
viruses (computer) 125
voice analysis 92
volcanos 19, 21
volunteers 23

W

Warrington bombing 9
water supply 163
weapons of mass destruction (WMD) 1
Web sites
    academic and institutions 209–210
    anthrax 213
    biological terrorism 211–212
    communications centers 207
    counterterrorism 210–211
    critical infrastructure protection
        212–213
    Federal Response Plan (FRP) 211
    government 208–209
    plague 214
    smallpox 213
    terrorism 207–208
white supremacists 9
wildland fires 18, 19, 53
winter storms 19, 20

wireless equipment 89
WMD (weapons of mass destruction) 1
    availability of 176
    events 1
        domestic preparedness 78–79
        expert information on 79–80
        psychological aspects of 188
    and facility security plans 164
    sources of 175
    and terrorism 174
    See also NBC (nuclear, biological, or
        chemical) agents/events
WordPerfect (software) 146
word-processing software 146
World Trade Center attacks (2001). See
    September 21, 2001 attacks
World Trade Center bombing (1993) 174

X

X-ray machines 90

Y

Yemen 9
Yersinia pestis 202
Yugoslavia 8